Hohenheim Tropical

Agricultural Series

SOIL CONSERVATION IN ANDEAN CROPPING SYSTEMS

Soil Erosion and Crop Productivity in Traditional and Forage-Legume Based Cassava Cropping Systems in the South Colombian Andes

by

Martin Ruppenthal

Margraf Verlag

This research was carried out in cooperation and with the support of the **Centro Internacional de Agricultura Tropical (CIAT)** in Cali, Colombia

Financial support was received from **Bundesministerium für Wirtschafliche Zusammenarbeit und Entwicklung (BMZE)** through a contract with **Deutsche Gesellschaft für Technische Zusammenarbeit (GTZ)**

CIP-Titelaufnahme der Deutschen Bibliothek

Ruppenthal, Martin:
Soil conservation in Andean cropping systems : soil erosion and crop productivity in traditional and forage-legume based cassava cropping systems in the South Colombian Andes / by Martin Ruppenthal. [Ed.: Center for Agriculture in the Tropics and Subtropics, University of Hohenheim. Managing ed.: Dietrich E. Leihner]. - Weikersheim : Margraf, 1995
(Hohenheim tropical agricultural series ; 3)
ISBN 3-8236-1248-4
NE: GT

Editor:
Center for Agriculture in the Tropics and Subtropics
University of Hohenheim
70593 Stuttgart
Germany

Managing Editor:
Dietrich E. Leihner
Director, Institute of Plant Production in the Tropics and Subtropics
University of Hohenheim

Assistance:
Thomas H. Hilger

Cover photo:
Lemon grass and vetiver grass barriers have shown excellent potential for soil conservation in mid-altitude Andean agriculture (Leihner).

Printing and binding:
F. & T. Müllerbader, Filderstadt
Germany

© Margraf Verlag, 1995
P.O. Box 105
97985 Weikersheim, Germany

ISBN 3-8236-1248-4
ISSN 0941-4894

CONTENTS

ACKNOWLEDGEMENTS

First of all, I am grateful to Professor Dr. D.E. Leihner, University of Hohenheim, under whose generous guidance I undertook this study.

At the **Centro Internacional de Agricultura Tropical (CIAT)**, I am indebted to Dr. M. El-Sharkawy, Dr. K. Müller-Sämann and Jesus Castillo F. for valuable discussions and practical help, to Dr. D. Laing and Prof. Dr. R. Schultze-Kraft for helpful advice on vetiver grass and forage legumes. I would also like to express my thanks to Margarita Pulgarin, for her friendly assistance in the office. I am grateful to D. Peña (field operations), O. Mosquera (soil laboratory), R. Narvaez (field operations Santander de Quilichao) and L.F. Cadavid (Santander de Quilichao). Special thanks are due to Luis E. Mina, Manuel Molina, Jose L. Adarve, Hector Mina, Juan Urrego, Nestor Caravali and Julio Ordoñez for their responsible field work and good company. My gratitude goes to the family of Guillermo and Rosa Helena Velasco in Mondomo, who provided their land to conduct part of this study.

At the **University of Hohenheim**, I am much obliged to Norbert Steinmüller for his statistical assistance and friendship, and to Dr. T. Hilger for helpful suggestions.

Financial support in Colombia and Germany was provided by the Bundesministerium für Wirtschaftliche Zusammenarbeit und Entwicklung (BMZE) through the Deutsche Gesellschaft für Technische Zusammenarbeit (GTZ) and in Germany by the Federal State of Baden-Württemberg. CIAT provided **logistical support** during the practical phase of this study (1990-1992).

PREFACE

Research on soil erosion and soil degradation processes has to be a long-term activity. This is because damage to agricultural lands from erosion does not normally become aparent in a single year and the loss of soil productivity is not instantaneous but rather of a gradually progressing nature. Therefore, solutions derived from a scientific examination of these processes can not be found overnight. It is for this reason that special effort has been put into the development of a sustained, long-term research programme in the tropical Andes which alone will provide answers to how soil erosion occurs and how it may best be prevented in that part of the world.

The research programme on which further information is provided in this report, is now in its 9th year and efforts to maintain scientific continuity and also to procure the necessary funding for it have not been small. Much insight has been gained into the factors governing soil erosion and soil degradation in the Andean region of Northern South America where relatively young soils (Inceptisols) prevail. The erosivity of tropical rainstorms in the region is now better understood although some of the basic information to actually calculate rainfall erosivity is still missing. Specific parameters of Inceptisols most accurately characterising their erodibility have been identified. Soil conservation strategies and technologies are emerging as a result of the cropping systems research continued in the project.

The first report summarising much of the basic information on erosion processes in Colombia's South Andean region has been published 3 years ago (Reining, 1992). Most appropriately, it was the first issue of a new series the Tropical Center of the University of Hohenheim is editing, the "Hohenheim Tropical Agricultural Series". It is now time to come back to the subject of soil erosion and soil conservation in the tropical Andes and report on further results and developments in this long-term research endeavour.

Hohenheim, May 1995

Dietrich E. Leihner
Managing Editor

SUMMARY

Worldwide, soils are lost to agricultural productive use at an alarming rate. Soil erosion by water and wind is estimated to be the single most important factor causing soil degradation. It is a much more serious problem in countries of the developing world than of temperate regions. Cassava (*Manihot esculenta* Crantz), an important subsistence and cash crop of small farmers, is grown up to 1,800 m a.m.s.l. in the tropical Andes, mainly as a sole crop on already degraded soils. Due to its slow initial development fields planted to cassava are especially prone to erosion. There is an urgent need to develop alternative cropping practices, adapted to local political and socioeconomic conditions, which maintain or even increase soil fertility and crop productivity.

The objectives of the present investigation, which is part of a broader research project on soil conservation in the South Colombian Andes (on station and on farm) were:
(i) to determine the erosivity of rainfall and the erodibility of soils, (ii) to evaluate the erosion control effectiveness of alternative practices in cassava based cropping systems and their productivity, with special emphasis on forage legume intercropping, (iii) to determine losses of organic matter and nutrients in sediments and water runoff and (iv) to investigate the changes in chemical and physical properties of soils, induced by erosion and management practices.

Erosion trials under natural rainfall were established in the years 1986-87 at two locations on uniform slopes with gradients from 7 to 20 %. In the beginning, effects of soil tillage, intercropping with grain legumes and the application of mulch were examined with regard to their soil conservation effectiveness and impact on crop productivity. The trials of the present study were started in May 1990 and continued during two cropping periods of cassava until April 1992. Three replications were established in *Santander de Quilichao* (hereafter termed "Quilichao") and two in *Mondomo*. The soils, both Inceptisols, are highly acid, rich in organic matter but low in available plant nutrients. Their physical characteristics such as infiltration and degree of aggregation are favorable.

The cropping systems under investigation were: (1) permanently clean tilled fallow, (2) cassava on contour ridges, (3) cassava on the flat, (4) cassava in association with *Pueraria phaseoloides* (Roxb.) Benth. (replaced by *Centrosema macrocarpum* Benth. in Mondomo),

(5) cassava with *Zornia glabra* Desv. and (6) cassava with *Centrosema acutifolium* Benth. (7) cassava, planted between contour strips of *Vetiveria zizanioides* (L.) Nash. and (8) cassava planted between contour strips of *Pennisetum purpureum* Schumach. Legumes were seeded simultaneously with cassava planting. In the second cropping period cassava was planted in minimum tillage in the legume sward. Fallow plots were 22.1 m long and 11 m wide, whereas all other plots measured 16 m x 8 m. The collection system for soil and runoff consisted of a collecting channel at the lower end of the plots, which was connected with a splitter and a tank to collect runoff water. At both locations monthly recording raingauges were established. Chemical and physical properties of topsoils were monitored over the two cropping periods as well as organic matter and nutrient contents in sediments and runoff. Plant growth and soil cover were measured regularly. Yields of cassava and legumes were determined at the end of the respective growth cycles.

The potential of soil erosion in the South Colombian Andes is high. Calculated erosivity factors (R) of the Universal Soil Loss Equation (USLE) in Quilichao were 9,042 MJ ha^{-1} mm h^{-1} a^{-1} in the first and 11,078 in the second cropping period, in Mondomo 7,548 and 5,142, respectively. In a preliminary approach, the USLE R-factor can be considered as a promising erosivity index. Coefficients of correlation with soil losses from permanent bare fallow plots ranged from 0.77 to 0.82 in Quilichao and from 0.85 to 0.90 in Mondomo ($P<0.001$). The inclusion of a simple term, to account for the influence of antecedent soil moisture on erosion, slightly improved correlations in Quilichao only. Over the two years study period about 40 % of the rain fell with intensities exceeding 40 mm h^{-1}. Average energy load of the rainfall was 21.5 J m^{-2} mm^{-1} in Quilichao and 20.9 J m^{-2} mm^{-1} in Mondomo. Precipitation was below average and suggests higher erosivity and soil losses in normal years. Two year losses from permanent bare fallow plots were about 3.8 cm of topsoil in Quilichao and 3.7 cm in Mondomo. Even on slopes of moderate gradients all fertile topsoil of permanent bare fallow plots may thus be lost within a decade.

Soil erodibility was found to be a dynamic property. It is lowest immediately after clearing of grassland or bush fallow. Erodibility measured on permanent clean tilled fallow plots under natural rainfall is classified as low to moderate in Quilichao and low in Mondomo. Five years of permanent bare fallowing lead to a steady increase of the K-factor (USLE) from 0.004 to 0.015 in Quilichao and from 0.003 to 0.012 in Mondomo (measured in SI-units). The nomograph of the USLE proved to be not applicable to estimate the erodibility of Inceptisols, such as those found in the South Colombian Andes as predicted values greatly underestimate measured K-values from continuously clean tilled fallow plots.

Besides rainfall characteristics and soil erodibility, soil cover and soil surface conditions at the time of rainfall were the most important factors governing the erosion process. In the first cropping period in Quilichao, soil losses were highest in the intercropping systems and increased with increasing degree of "human traffic" for resowing forage legumes which were difficult to establish. An amount of 27.2 t ha^{-1} were lost with *Z.glabra*, 17.0 t ha^{-1} with *C.acutifolium* and 14.2 t ha^{-1} with *P.phaseoloides*. One heavy rainstorm shortly after planting caused 70 to 80 % of all losses. Soils in Mondomo did not react as sensitive to soil disturbance. Losses were highest in traditionally flat cropped cassava with 4.7 t ha^{-1}. Well established and uniformly distributed forage legumes, dense, gapless stands of contour grass barriers and contour ridging on these flat slopes effectively controlled soil erosion. *V.zizanioides* as a contour barrier proved to be a very promising method of erosion control. Runoff rates were low and ranged from 2.2 to 8.2 % of annual precipitation for the cassava cropping systems. Over 70 % of the total loss of available Mg, K and P was associated with sediments on bare fallow plots. Proportion of runoff bound nutrients increased when soil loss decreased. In cassava cropping systems amount of K, lost by runoff water, always exceeded losses by sediments. Probably interflow water led to an enrichment of K and Mg in runoff water. Concentrations of K and Mg were higher in runoff water from cassava treatments than from bare fallow. P concentrations, however, were higher from the bare plots. Eroded hillsides in the study region may thus contribute to the eutrophication of lakes and rivers.

The selectivity of the erosion process can be expressed by the enrichment ratio, which is the ratio of the concentration of a component in sediments to that of the corresponding matrix soil. Enrichment ratios were low on these highly aggregated soils. In Quilichao ratios were found to be in the following order: Organic matter (0.93 - 0.97), total N (0.96 - 0.99), exchangeable Mg (1.04 - 1.12), Bray-II P (1.07 - 1.25) and exchangeable K (1.20 - 1.30). Slightly higher ratios for organic matter and total N were found in Mondomo.

Generally greatest cassava yields were achieved on contour ridges or flat cropped, ranging from 22.7 to 35.7 t ha^{-1} in Quilichao and from 13.4 to 18.4 t·ha^{-1} in Mondomo. Yield reductions by forage legumes were slight in the establishment year, but strong in the following cropping period (> 40 %). Dry matter production of legumes ranged from 4.7 to 6.7 t ha^{-1} in Quilichao and from 1.5 to 5.5 t ha^{-1} in Mondomo in both cropping periods.

Differences in concentrations of organic matter and nutrients in topsoils among cassava cropping systems generally were not significant. Lack of significance is due to the relatively short time period of the investigation and the way the cropping systems were managed. A smaller nutrient extraction by low-yielding cassava in the legume-cassava intercropping systems was offset by the additional nutrient removal of the forage legumes in intercropping systems (cut and carry system). The time of cropping had the greatest effect on chemical soil properties. Annual fertilization increased levels of exchangeable Ca, Mg and of Bray-II P. Free Al-content decreased. Organic matter content declined in Mondomo, but increased in Quilichao.
Aggregate stability, measured by a modified YODER-wet sieving procedure, was found to be higher in Quilichao than in Mondomo and does not reflect the differences in soil erodibility found between both locations. Lowest aggregate stability among cassava cropping systems was found in flat and intercropped cassava. Intense drying and wetting cycles seem to have enhanced stability of soil aggregates in contour ridges. It came near old grassland and bush fallow soils and may be one reason for the excellent cassava yields on ridged soil.
Sand and silt content of the trial plots increased during the two-year study period, whereas clay content decreased. Sediments showed higher sand contents than the original soils.
Final infiltration rates of cropped plots were higher in Mondomo with 15.2 cm h^{-1} than in Quilichao with 3.2 cm h^{-1}. At the end of the study period significant differences among treatments were found for the early stages of infiltration. Cassava associated with *C.acutifolium* in Quilichao, with *C.acutifolium* and *Z.glabra* in Mondomo showed a higher permeability than bare fallow plots. In Quilichao, initial rates were lowest, compared to other cropping systems when cassava was intercropped with *Z.glabra*. This may be the reason for the higher runoff rates in this treatment and confirm the higher susceptibility of soils in Quilichao to mechanical stress (compaction, aggregate desintegration by walking).
Bulk densities of cropped topsoils were slightly higher in Quilichao than in Mondomo. One year after establishing the present trial no significant differences among treatments occurred for the moisture retained at high ranges of matrix potential (-1 and -6 kPa). At wilting point soils under cassava with contour barriers of *P.purpureum* and *V.zizanioides* showed a tendency to higher moisture retention than soils of the intercropping treatments. Available water capacity between -30 and -1500 kPa was highest in heavily eroded bare fallow soils (9.2 Vol.%). Limited data at the end of the study period indicated a significant increase in macroporosity at a potential of -1 kPa in cassava associated with *C.acutifolium*.

Contour barriers of dense, gapless strips of grasses (also when combined with forage legumes) or contour ridging on flat to moderate slopes effectively control erosion in cassava cropping on Inceptisols in the South Colombian Andes and produce reasonable crop yields. Similar results were obtained by Reining (1992) in previous years at the two study locations. Forage legume intercropping cannot be a recommended practice for the smallholder in the study region. Inceptisols in the montainous regions of South West Colombia are of low soil fertility, but of an excellent physical stability. For this reason, eroded hillsides still posess the potential of agricultural production once soil fertility constraints can be alleviated.

RESUMEN

EROSION DE LOS SUELOS Y PRODUCTIVIDAD DE LOS SISTEMAS DE PRODUCCION DE YUCA TRADICIONALES Y ALTERNATIVAS EN LA ZONA ANDINA DEL SUR OESTE COLOMBIANO

Mundialmente el recurso natural del suelo se pierde en forma alarmante para el uso agrícola. La erosión, ocasionada por el agua y el viento, es considerada como el factor individual más importante de la degradación de los suelos. El problema de la erosión es más importante en el trópico que en las latitudes altas. La yuca (*Manihot esculenta* Crantz) es un cultivo importante comercial y de autoabastecimiento de los pequeños agricultores. En general es cultivada en los Andes tropicales en suelos erodados como monocultivo hasta una altura de 1800 m s.n.m.. Pero debido a su lento desarrollo inicial el riesgo de la erosión en los cultivos es alto. La búsqueda de sistemas de producción alternativos adaptados a las condiciones locales socioeconómicas y políticas es de importancia especial para la investigación. Ademas, es necesario que estos sistemas mantengan, ó en lo possible, mejoren la fertilidad y la productividad de los suelos. Los experimentos del presente estudio forman parte de un proyecto de investigación sobre conservación de suelos en los Andes del Sur Oeste Colombiano, que tiene los siguientes objetivos: 1) Determinar la erosividad de la precipitación y la erodabilidad de los suelos, 2) evaluar la productividad y la eficiencia de las medidas de control de erosión en sistemas de producción de yuca tradicionales y alternativos, en particular considerando asociaciones con leguminosas forrajeras, 3) determinar las pérdidas de materia orgánica y de nutrientes en los sedimentos y en el agua de escorrentía, 4) investigar el impacto de la erosión de agua y del manejo del cultivo a propiedades químicas y físicas del suelo.

Ensayos de erosión, sobre pendientes uniformes, que oscilaban entre 7 al 20 %, fueron instalados en dos localidades del Departamento del Cauca, en los años 1986-87. Hasta Abril de 1990 se investigó diferentes sistemas de producción de yuca, incluyendo formas de la cultivación del suelo, la asociación con leguminosas de grano y la aplicación de mulch. Los ensayos del presente estudio se iniciaron en Mayo 1990 y terminaron después de dos cíclos de producción de yuca en April 1992. Se utilizaron tres repeticiones en Santander de Quilichao y dos en Mondomo. Los suelos altamente ácidos, ambos Inceptisoles, tienen altos contenidos de materia orgánica, pero se caracterizan por baja fertilidad. En cambio, tienen propiedades físicas favorables tales como una infiltrabilidad y un grado de la agregación altos.

En el presente estudio fueron evaluados los siguientes tratamientos: 1) suelo desnudo, mantenido libre de vegetación desde 1986-87, 2) yuca en caballones de contorno, 3) yuca tradicional en monocultivo, 4) yuca asociada con *Pueraria phaseoloides* (Roxb.) Benth., (remplazada por *Centrosema macrocarpum* Benth. en Mondomo), 5) con *Zornia glabra* Desv. 6) con *Centrosema acutifolium* Benth., 7) yuca con barreras en contorno de *Vetiveria zizanioides* (L.) y 8) de *Pennisetum purpureum* Schuhmach.. Las leguminosas fueron sembradas simultaneamente con la yuca. En el segundo cíclo la yuca se sembró en mínima labranza dentro de la leguminosa establecida. Las parcelas con yuca tenían una dimensión de 16 m x 8 m, las parcelas libres de vegetación de 22.1 m x 11 m. El sistema de colección de sedimentos y del agua de la escorrentía consistió en una canaleta en la parte inferior de la parcela, la cual fue conectada con un distribuidor y un tanque colector. En cada localidad se instaló un pluviógrafo de registro mensual. Las propiedades químicas y físicas de los suelos, como tambien los contenidos de materia orgánica y nutrientes en los sedimentos y las aguas de escorrentía fueron analizados durante los dos cíclos de producción de yuca. El desarrollo de las plantas y la cobertura del suelo se determinó en intervalos periódicos.

El potencial de erosión en los Andes del Sur Oeste Colombiano es alto. Los factores de erosividad (R) de la "Universal Soil Loss Equation" (USLE, Wischmeier y Smith, 1978) fueron 9042 (MJ ha^{-1} mm h^{-1}) en el primer y 11078 en el segundo cíclo en Quilichao, 7548 y 5142 en Mondomo, respectivamente. Los coeficientes de correlación entre las pérdidas de suelo del tratamiento 1 (suelo desnudo) y los factores R del la USLE variaron entre 0.77 y 0.82 en Quilichao, y entre 0.85 y 0.90 en Mondomo. Considerando el corto lapso del

tiempo de la presente investigación, se puede considerar R como un índice promisorio de la erosividad.
Incluyendo un simple parámetro para la influencia de la humedad del suelo a la erosión, las correlaciones fueron mejoradas solo en Quilichao. En Mondomo, las precipitaciones se encontraron considerablemente por debajo del promedio anual, sugiriendo una más alta erosividad y pérdida de suelos en años normales. Aproximadamente el 40 % de la precipitación excedió una intensidad de 25 mm h^{-1} en los dos años de la presente investigación. La energía kinética promedia de la lluvia fue de 21.5 J m^{-2} mm^{-1} en Quilichao y de 20.9 J m^{-2} mm^{-1} en Mondomo. En estos dos años 3.8 cm de suelo se perdieron en el tratamiento 1 (suelo desnudo) en Quilichao y 3.7 cm en Mondomo. Por lo tanto, el suelo fértil superficial se puede perder en una sola década.

La erodabilidad es una propiedad dinámica: Immediatamente después de la cultivación de pastos o barbechos la suceptibilidad de los suelos a la erosión es muy baja. La erodabilidad (factor K de la USLE) aumentó en cinco años, teniendo el suelo libre de vegetación, desde 0.004 hasta 0.015 t ha h ha^{-1} MJ^{-1} mm^{-1} en Quilichao y desde 0.003 hasta 0.012 en Mondomo. Hasta la presente la erodabilidad, determinado de la pérdida de suelos de parcelas desnudas dividido por el correspondiente factor R de la USLE (método directo), es clasificada como baja. El USLE nomograph (método indirecto) no es apto para determinar la erodibilidad de Inceptisols en la region estudiada, pues los valores de K subestimaron considerablemente la erodabilidad, determinado con el método directo.

Además de las propiedades físicas de la precipitación (cantidad, distribución, erosividad) y la erodabilidad de los suelos, los factores más importantes para determinar la erosión son el grado de la cobertura del suelo y la condición de la capa superficial del suelo en el momento del evento erosivo. En Quilichao en el primer cíclo del cultivo de yuca las pérdidas de suelo más altas ocurieron en las asociaciones con leguminosas forrajeras. Las pérdidas ascendieron a medida que aumentó el disturbio del suelo para establecer las leguminosas, lo cual fue muy difícil. Una cantidad de 14.2 t ha^{-1} se perdió con *P. phaseoloides*, 17.0 t ha^{-1} con *C. acutifolium* y 27.2 t ha^{-1} con *Z. glabra*. Un fuerte evento de lluvia, inmediatamente después de la siembra, causó del 70 al 80 % de la pérdida anual total. Los suelos en Mondomo reaccionaron con menos susceptibilidad al disturbio adicional del suelo causado por el establecimiento de las leguminosas. Las pérdidas más altas ascendieron a 4.7 t ha^{-1} registradas en la yuca cultivada tradicionalmente. Las medidas más eficientes para controlar la erosión fueron las leguminosas bien establecidas y

uniformemente distribuidas, barreras densas y uniformes de pasto en contorno y la cultivación de yuca en caballones en contorno en pendientes bajas y moderadas. Como medida muy promisoria se presentó el uso de *Vetiver zizanioides* como barrera en contorno. Las porciones de lluvia, pérdidas como agua de escorrentía, fueron bajas con rangos entre 2.2 % y 8.2 % de la precipitación en los sistemas de producción de yuca.
Más del 70 % de la pérdida total de Mg, K y P disponibles en el tratamiento 1 (suelo libre de vegetación) es asociado con los sedimentos. La proporción de nutrientes, que se perdieron con el agua de escorrentía, se incrementó con la disminución de la pérdida de suelos. Las pérdidas de K en la escorrentía excedieron las pérdidas en los sedimentos en todos los sistemas de producción de yuca. Tal vez el desagüe sub-superficial (interflow) llevó a un enriquecimiento en K y Mg en la escorrentía. Las concentraciones de K y Mg en la escorrentía de los sistemas de yuca fueron más altas y de P más bajas que en la escorrentía de los suelos desnudos. Por lo tanto los suelos erodados de la región pueden contribuir sustancialmente a la eutroficación de las aguas.

Debido a la buena agregación de los suelos, las tasas del enriquecimiento (TE) de materia orgánica y nutrientes en los sedimentos fueron bajas. Las siguientes TE's se reportaron para Quilichao: Materia orgánica (0.93 - 0.97), N total (0.96 - 0.99), Mg disponible (1.04 - 1.12), Bray-II P (1.07 - 1.25) y K disponible (1.20 - 1.30). TE's un poco más altas se encontraron en Mondomo para materia orgánica (1.03) y N total (1.03 - 1.14).

La producción de yuca más alta se obtuvo con el sistema tradicional y con caballones en contorno. Entre 22.7 y 35.7 t ha^{-1} de raices frescas fueron cosechadas en Quilichao y entre 13.4 y 18.4 t ha^{-1} en Mondomo. En el año de establecimiento de las leguminosas la producción de yuca se redujó ligeramente, pero fuertemente en el siguiente período (> 40 %). En ambos períodos se cosecharon entre 4.7 y 6.7 t ha^{-1} de materia seca de leguminosas en Quilichao y entre 1.5 y 5.5 t ha^{-1} en Mondomo.

En general, al final del período del ensayo entre los diferentes sistemas de producción no se presentaron diferencias significativas en las concentraciones de materia orgánica y nutrientes en la capa superficial de los suelos (0-20 cm). La falta de significancia se atribuye al lapso corto del tiempo de la presente investigación y al modo como se ejecutaron los sistemas de producción. Una reducida extracción de nutrientes por la yuca en los sistemas asociados con leguminosas fue compensada por una extracción adicional por las leguminosas forrajeras, las cuales fueron cortadas regularmente y el material quitado de las parcelas. El número de los años de producción y la fertilización anual para la yuca ejerció el efecto más grande a la fertilidad de los suelos. Las concentraciones de Ca y Mg intercambiables y de Bray-II P aumentaron y las de Al libre disminuyeron. En Mondomo la materia orgánica fue reducida continuamente con el creciente número de años en cultivación, mientras que en Quilichao en el lapso del tiempo de la presente investigación el contenido de materia orgánica aumentó.

La estabilidad de agregados al agua, determinada con un método modificado de YODER, fue más alta en Quilichao que en Mondomo. Estos resultados contrastaron con los valores de la erodabilidad de los suelos. Entre los sistemas de producción de yuca las estabilidades fueron más bajas en la yuca tradicionalmente cultivada y asociada con leguminosas forrajeras. Los agregados de los caballones en contorno mostraron estabilidades extraordinarias, probablemente causado por los frecuentes cambios entre fases secas y húmedas en los caballones. Los valores alcanzaron casi el nivel de agregados bajo barbecho natural o pastos permanentes. Esta podría ser una razón por los buenos rendimientos de yuca en este tratamiento.
Los contenidos de arenas y de limos de los suelos aumentaron desde 1990 hasta 1992, en cambio se redujeron las arcillas. Los sedimentos mostraron contenidos más altos de arena que los suelos de donde se originaron.
La tasas de infiltración en equilibrio (método de dos anillos concéntricos) fueron más altos en Mondomo con 15.2 cm h^{-1} que en Quilichao con 3.2 cm h^{-1}. Al final del periodo de la investigación se pudieron observar diferencias significativas entre los tratamientos solamente en la fase inicial del proceso de la infiltración: Se encontró una mayor permeabilidad de agua en los suelos con yuca y *C.acutifolium* en Quilichao y con *C.acutifolium* y *Z.glabra* en Mondomo en comparación con los suelos desnudos. Las tasas de la infiltración inicial en suelos de la asociación yuca-*Z.glabra* en Quilichao fueron más bajas que en otros sistemas de producción. Esto explica las cantidades más altas de la escorrentía de agua en este tratamiento y confirma la mayor suceptibilidad de los suelos en Quilichao a la compactación par pisoteo.
Las densidades aparentes de las parcelas cultivadas con yuca fueron mayores en Quilichao que en Mondomo. Doce meses después del establecimiento del presente ensayo no se encontraron diferencias significativas entre los tratamientos en los contenidos de agua de la capa superficial en potenciales (matrix potentials) altas (-1 y -6 kPa), pero en potentiales menores. Por todos los potenciales los contenidos de agua fueron más bajas en los suelos de las asociaciones y más altas, cuando se usaron barreras de *P.purpureum* en contorno para el

control de la erosión en yuca. Los contenidos de agua disponible entre -30 y -1500 kPa variaron entre 5.5 y 9.2 Vol.%. Las capas superficiales de los suelos desnudos mostraron los valores más altos con diferencias significativas al sistema tradicional de yuca. Mediciones, conducidos dos años despues del inicio del presente estudio, mostraron un significativo aumento de la macroporosidad en los suelos con yuca y *C.acutifolium*.

Yuca asociada con leguminosas no presenta una alternativa a la producción tradicional para controlar la erosión y obtener rendimientos altos. Cuando se alivia la baja fertilidad con fertilizantes químicos y orgánicos, aún suelos erodados tienen la capacidad de producción agrícola gracias a sus propiedades físicas favorables.

ZUSAMMENFASSUNG

BODENEROSION UND PRODUKTIVITÄT VON TRADITIONELLEN UND ALTERNATIVEN KLEINBÄUERLICHEN MANIOKANBAU-SYSTEMEN IN DEN ANDEN SÜDWESTKOLUMBIENS

Weltweit geht die Ressource Boden in alarmierendem Maße für die landwirtschaftliche Nutzung verloren. Erosion durch Wasser und Wind wird als die wichtigste Einzelursache für die Bodenzerstörung angesehen. Das Problem der Bodenerosion ist in den Tropen weitaus größer als in den gemäßigten Breiten. Maniok (*Manihot esculenta* Crantz), eine wichtige Subsistenz- und Marktfrucht von Kleinbauern, wird in den tropischen Anden hauptsächlich in Reinkultur auf zumeist erodierten Böden bis auf etwa 1800 m N.N. angepflanzt. Aufgrund seiner langsamen Jugendentwicklung ist das Erosionsrisiko im Maniokanbau jedoch hoch. Die Forschung zur Entwicklung alternativer Anbausysteme, die zum einen an die lokalen politischen und sozioökonomischen Bedingungen angepaßt sind und zum anderen die Produktivität und Bodenfruchtbarkeit erhalten und wenn möglich verbessern, ist von besonderer Wichtigkeit für die Erhaltung annehmbarer Lebens- und Einkommensbedingungen in tropischen Bergregionen.

Die vorliegenden Untersuchungen, die in ein größeres Forschungsprojekt über Bodenschutz in den Anden Südwest Kolumbiens integriert waren, hatten folgende Zielsetzung: 1) die Bestimmung der Erosivität der Niederschläge und der Erodibilität der Böden, 2) die Evaluierung der Wirksamkeit von Maßnahmen zur Erosionskontrolle in traditionellen und alternativen Maniokanbausystemen und deren Produktivität, unter besonderer Berücksichtigung des Mischanbaus mit Futterleguminosen, 3) die Bestimmung der Verluste an organischer Substanz und Nährstoffen mit Sedimenten und Oberflächenabfluß und 4) die Untersuchung der Auswirkungen von Wassererosion und Anbausystem auf chemische und physikalische Bodeneigenschaften.

Auf zwei Standorten im Departamento Cauca wurden in den Jahren 1986/1987 Erosionsversuche auf gleichförmigen Hängen mit Neigungen von 7 bis 20 % angelegt. Die Bodenbearbeitung, der Mischanbau mit Körnerleguminosen und die Auswirkungen von Mulchanwendung im Maniokanbau standen im Vordergrund dieser Untersuchungen. Die Versuche zur vorliegenden Arbeit begannen im Mai 1990 und endeten im April 1992 nach zwei Maniokanbauperioden. Sie bestanden aus drei Wiederholungen in *Santander de Quilichao* (hiernach "Quilichao" genannt) und zwei in *Mondomo*. Die stark sauren Böden, beide Inceptisols, haben hohe Gehalte an organischer Substanz, weisen jedoch eine geringe chemische Bodenfruchtbarkeit auf. Andererseits sind die physikalischen Eigenschaften (wie Wasserdurchlässigkeit, Aggregierungsgrad) als günstig einzustufen.

Folgende Anbausysteme wurden untersucht: 1) Langjährige Schwarzbrache, 2) Maniok auf Konturdämmen, 3) Maniok auf flach gepflügtem Land (traditioneller Anbau), 4) Maniok im Mischanbau mit *Pueraria phaseoloides* (Roxb.) Benth. (ersetzt durch *Centrosema macrocarpum* Benth. in Mondomo), 5) Maniok mit *Zornia glabra* Desv. und 6) Maniok mit *Centrosema acutifolium* Benth., 7) Maniok mit Konturstreifen von *Vetiveria zizanioides* (L.) und 8) Maniok mit Konturstreifen von *Pennisetum purpureum* Schuhmach.. Die Leguminosen wurden gleichzeitig mit der Maniokpflanzung ausgesät. In der zweiten Anbauperiode wurde Maniok in Minimalbodenbearbeitung in den bestehenden Leguminosenbestand gepflanzt. Die Versuchsparzellen mit Maniok waren 16 m lang und 8 m breit, die Schwarzbracheparzellen 22.1 m lang und 11 m breit. Das Auffangsystem für Sedimente und Oberflächenabfluß bestand aus einem Kanal am unteren Ende der Parzellen, der mit einem Verteiler und einem Tank verbunden war. An beiden Standorten waren selbstregistrierende Regenschreiber installiert. Chemische und physikalische Bodeneigenschaften, sowie Gehalte an organischer Substanz und Nährstoffen in Sedimenten und Abfluß, wurden über beide Anbauperioden untersucht. Pflanzenwachstum und Bodenbedeckung wurden in regelmäßigen Abständen bestimmt.

Das Erosionspotential in den südkolumbianischen Anden ist hoch. Berechnete Erosivitätsfaktoren (R) der "Universal Soil Loss Equation" (USLE, Wischmeier und Smith,

1978) in Quilichao lagen bei 9042 (MJ ha^{-1} mm h^{-1} a^{-1}) in der ersten und bei 11078 in der zweiten Anbauperiode, in Mondomo bei 7548 und 5142. Die Korrelationskoeffizienten der Bodenabträge von langjähriger Schwarzbrache mit den USLE R-Faktoren lagen zwischen 0.77 und 0.82 in Quilichao und zwischen 0.85 und 0.90 in Mondomo. Somit kann der R-Faktor vorläufig, unter Berücksichtigung des kurzen Zeitraumes der vorliegenden Untersuchung, als annehmbarer Erosivitätsindex angesehen werden. Durch die Einbeziehung eines einfachen Terms für den Einfluß der Bodenfeuchte auf das Erosionsgeschehen, wurden die Korrelationen nur in Quilichao leicht verbessert. Die Niederschläge lagen zum Teil beträchtlich (Mondomo) unter dem langjährigen Mittel und lassen eine höhere Erosivität und höhere Bodenverluste in 'normalen' Jahren erwarten. In der zweijährigen Versuchsperiode fiel etwa 40 % des Regens mit Intensitäten größer als 25 mm h^{-1}. Die durchschnittliche kinetische Energie des Niederschlages betrug 21.5 J m^{-2} mm^{-1} in Quilichao und 20.9 J m^{-2} mm^{-1} in Mondomo. In zwei Jahren wurden 3.8 cm Boden von Schwarzbracheparzellen in Quilichao und 3.7 cm in Mondomo abgetragen. Der gesamte fruchtbare Oberboden kann in nur einem Jahrzehnt völlig verloren gehen.

Die Erodibilität erwies sich als eine dynamische Bodeneigenschaft. Unmittelbar nach dem Umbruch von Grasland und Buschbrache war sie am niedrigsten. Die Empfindlichkeit der Böden gegenüber Erosion stieg innerhalb von fünf Jahren unter Schwarzbrache von 0.004 auf 0.015 (t ha h ha^{-1} MJ^{-1} mm^{-1}) in Quilichao und von 0.003 auf 0.012 in Mondomo. Die Erodibilität, berechnet als das Verhältnis zwischen dem Bodenabtrag von Schwarzbrache und dem USLE R-Faktor, wird bis dato als niedrig bis mäßig in Quilichao und als niedrig in Mondomo eingestuft. Der USLE Nomograph erscheint nicht geeignet, die Erodibilität von Inceptisols in der Versuchsregion mit einiger Genauigkeit vorrauszuberechnen. Mit dem Nomograph berechnete K-Werte unterschätzten die auf Schwarzbrache gemessenen Werte stark.

Neben den Eigenschaften der Niederschläge (Menge, Erosivität und Verteilung) und der Erodibilität der Böden sind der Grad der Bodenbedeckung und der Zustand der Bodenoberfläche beim Auftreten eines Regenereignisses die wichtigsten Faktoren, die das Erosionsgeschehen bestimmen. In der ersten Anbauperiode in Quilichao traten die höchsten Bodenverluste in den Maniok-Mischanbausystemen auf. Sie nahmen mit zunehmender Bodenbelastung durch Betreten der Felder infolge mehrfach notwendiger Wiederaussaat der Leguminosen zu. Eine Bodenmenge von 27.2 t ha^{-1} wurde abgetragen beim Anbau mit

Z.glabra, 17.0 t ha^{-1} mit *C.acutifolium* und 14.2 t ha^{-1} mit *P.phaseoloides*. Ein einzelner Starkregen kurz nach der Pflanzung verursachte 70 bis 80 % des jährlichen Gesamtverlustes. Die Böden in Mondomo reagierten weniger empfindlich auf eine durch die Leguminosenetablierung bedingte zusätzliche Bodenbelastung. Die Bodenabträge waren mit 4.7 t ha^{-1} am höchsten in traditionell angebautem Maniok auf flach gepflügtem Land. Gut etablierte und gleichmäßig verteilte Futterleguminosen, dichte und lückenlose Graskonturstreifen und der Anbau mit Konturdämmen auf Hängen geringer bis mäßiger Neigung kontrollierten die Erosion am besten. Als vielversprechende Maßnahme des Bodenschutzes erwies sich der Einsatz von *V.zizaniodes* als Konturbarriere.

Die Raten des Oberflächenabflußes waren gering und lagen zwischen 2.2 und 8.2 % des Niederschlages in den Maniokanbausystemen.

Über 70 % des Gesamtverlustes von pflanzenverfügbarem Mg, K und P von Schwarzbrache waren mit den Sedimenten assoziiert. Der Anteil an Nährstoffen, die mit dem Oberflächenabfluß verloren gingen, nahm mit der Abnahme des Bodenabtrages zu. Die Verluste an K im Abfluß aller Maniokanbausysteme übertrafen diejenigen, die mit den Sedimenten verloren gingen. Vielleicht führte oberflächennaher Abfluß (interflow) zu einer Anreicherung von K und Mg im Oberflächenabfluß. Die Gehalte an K und Mg waren höher im Abfluß von Maniokanbausystemen als von Schwarzbrache. P-Gehalte waren hingegen höher im Abfluß von Schwarzbrache. Erodierte Hänge der Versuchsregion können somit wesentlich zur Eutrophierung der Gewässer beitragen.

Wie zu erwarten war die Anreicherung von organischer Substanz und Nährstoffen in Sedimenten von diesen gut aggregierten Böden gering. Die Anreicherungsraten waren in Quilichao für organische Substanz (0.93 - 0.97), Gesamt N (0.96 - 0.99), austauschbares Mg (1.04 - 1.12), Bray-II P (1.07 - 1.25) und für austauschbares K (1.20 - 1.30). Leicht höhere Raten wurden für organische Substanz und Gesamt N in Mondomo gefunden.

Beste Maniokerträge wurden beim Anbau auf flachem Land und auf Konturdämmen erzielt. Sie lagen zwischen 22.7 und 35.7 t ha^{-1} in Quilichao und zwischen 13.4 und 18.4 t ha^{-1} in Mondomo. Im Etablierungsjahr der Futterleguminosen gingen die Maniokerträge nur leicht, in der folgenden Anbauperiode aber stark zurück (>40 %). Die Trockenmasseerträge der Leguminosen in beiden Anbauperioden lagen zwischen 4.7 und 6.7 t ha^{-1} in Quilichao und zwischen 1.5 und 5.5 t ha^{-1} in Mondomo.

Am Ende der Versuchsperiode traten im allgemeinen zwischen den Maniokanbausystemen keine signifikanten Unterschiede in den Gehalten an organischer

Substanz und Nährstoffen der Oberböden auf. Dieser Mangel an Signifikanz wird auf die kurze Zeitspanne der vorliegenden Untersuchung und die Art, wie die Anbausysteme durchgeführt wurden, zurückgeführt. Ein niedrigerer Nährstoffentzug durch geringere Maniokerträge in den Mischanbausystemen wird durch den zusätzlichen Nährstoffentzug der Futterleguminosen ausgeglichen. Die Leguminosen wurden regelmäßig geschnitten und von den Feldern entfernt. Die Anzahl der Anbaujahre und damit verbunden die jährliche Düngung zu Maniok übte den größten Effekt auf die Bodenfruchtbarkeit aus. Die Gehalte an austauschbarem Ca, Mg und an Bray-II P stiegen an, während die Gehalte an freiem Al abnahmen. Die organische Substanz wurde in Mondomo mit zunehmender Anzahl der Anbaujahre kontinuierlich abgebaut, wohingegen es in Quilichao in den beiden Anbauperioden der gegenwärtigen Untersuchung zu einer Zunahme kam.

Die Stabilität der Aggregate, die mit einer modifizierten Naßsiebungsmethode nach YODER bestimmt wurde, war in Quilichao höher als in Mondomo. Diese Ergebnisse stehen im Gegensatz zu den Werten für die Erodibilität der Böden. Unter den Maniokanbausystemen zeigten Böden, die traditionell oder zusammen mit Futterleguminosen bebaut wurden, die niedrigsten Stabilitäten. Aggregate von Konturdämmen waren außerordentlich stabil. Die Werte erreichten fast das Niveau von altem Grasland oder Buschbrache, wofür der häufige Wechsel von Trocken- und Feuchtephasen in den Dämmen verantwortlich zu sein scheint. Unter anderem mag das ein Grund für die guten Maniokerträge dieser Behandlung sein.
Die Sand- und Schluffgehalte der Parzellen nahmen von 1990 bis 1992 zu, die Tongehalte hingegen ab. Sedimente wiesen einen höheren Anteil an Sand als der Ursprungsboden auf.
Die konstanten Infiltrationsraten (Doppelring-Methode) waren in Mondomo mit 15.2 cm h^{-1} höher als in Quilichao mit 3.2 cm h^{-1}. Am Ende der Versuchsperiode wurden signifikante Unterschiede zwischen den einzelnen Behandlungen nur im Anfangsstadium des Infiltrationsprozesses beobachtet: Die Wasserdurchlässigkeit von Böden mit Maniok und *C.acutifolium* in Quilichao und mit *C.acutifolium* und *Z.glabra* in Mondomo war signifikant höher als von Schwarzbrache. In Quilichao waren die anfänglichen Infiltrationsraten in der Maniok-*Z.glabra* Assoziation jedoch niedriger als in den anderen Anbausystemen. Dies erklärt die höheren Mengen an Oberflächenabfluß in dieser Behandlung und bestätigt die größere Empfindlichkeit der Böden in Quilichao gegenüber mechanischem Streß.

Die Lagerungsdichten der mit Maniok bebauten Flächen hingegen war höher in Quilichao als in Mondomo. Zwölf Monate nach Anlage des vorliegenden Versuches traten zwischen den Behandlungen keine signifikanten Unterschiede der Wassergehalte in den Oberböden bei hohen (-1 und -6 kPa), jedoch bei den niedrigeren Matrixpotentialen auf. Über alle Potentiale hinweg waren die Wassergehalte in den Böden von Mischanbausystemen am niedrigsten und am höchsten, wenn Konturstreifen aus *P.purpureum* zur Erosionskontrolle benutzt wurden. Die Gehalte an pflanzenverfügbarem Wasser zwischen pF 2.5 und 4.2 lagen zwischen 5.5 und 9.2 Vol.%. Höchste Werte wiesen die Oberböden von Schwarzbrache auf mit signifikanten Unterschieden zum traditionellen Anbausystem und zu Konturdämmen. Messungen, zwei Jahre nach Versuchsbeginn ausgeführt, ergaben einen signifikanten Anstieg der Makroporosität in Böden mit Maniok und *C.acutifolium*.

Verglichen mit dem in der Region üblichen, traditionellen Maniokanbau scheint der Mischanbau mit Futterleguminosen keine besonders vielversprechende Alternative zur Erzielung hoher Erträge bei gleichzeitiger Erosionskontrolle zu sein. Bei dieser Einschätzung muß allerdings die Vorläufigkeit dieser Ergebnisse und die geringe Zahl der zusammen mit Maniok getesteten Leguminosen berücksichtigt werden. Wenn der geringen Fruchtbarkeit der Böden mittels organischer und chemischer Düngung abgeholfen werden kann, sind selbst erodierte Standorte dank ihrer günstigen physikalischen Bodeneigenschaften zur landwirtschaftlichen Produktion befähigt.

ABBREVIATIONS

R	Rainfall erosivity factor of the USLE
K	Soil erodibility factor of the USLE
L	Slope length factor of the USLE
S	Slope gradient factor of the USLE
C	Crop management factor of the USLE
P	Conservation practice factor of the USLE
USLE	Universal Soil Loss Equation
A	Rainfall amount
E_u, KE	Kinetic energy
I	Rainfall intensity
I_{MAX}	Maximum intensity
P<0.1,+	Error probability level of 10 %
P<0.05,*	Error probability level of 5 %
P<0.01,**	Error probability level of 1 %
P<0.001,***	Error probability level of 0.1 %
ns	not significant
r	Coefficient of correlation
r^2	Coefficient of determination
TRT	Treatment
LOC	Location
CP	Cropping period
DM	Dry matter
ER	Enrichment ratios
OM	Organic matter

Part 1 *INTRODUCTION*

Approximately half of the world's total potentially productive land area of about 3.3 billion ha is already under cultivation. However, much of the remaining land is either marginal, or social, economic, political and ecological contraints limit its availability for development (El-Swaify, 1991). It has been estimated, that about 10 ha of arable land is lost to productive use every minute (5.3 million ha annually), of which 5 ha are through soil erosion, three through salinization, one through other degradation processes and one because of urbanization (El-Swaify et al., 1983).
The problem of soil erosion is much more serious in tropical than in temperate regions. The erosive nature of tropical rainfall and the increasing cultivation of marginal and steep lands has reduced the depth of fertile topsoil. Under these conditions, high value cash crops such as maize and beans can no longer be grown and less exigent, but low value crops such as cassava occupy their place. With its slow early growth and poor initial soil cover, cassava creates conditions favorable to water erosion and soil degradation, in particular when cultivated without fertilization. Thus, agricultural lands degrade further and environmental constraints build up to such a degree that it may be increasingly impossible for small scale hillside farmers to grow even cassava, their "crop of last resort" in an agronomically and economically successful way.

The present study was therefore carried out with the following objectives: (i) to gather basic information on rainfall erosivity and erodibility of tropical Inceptisols in the South Colombian Andes; (ii) to measure erosion losses (soil, runoff, organic matter and nutrients); (iii) to determine productivity of traditional and alternative cassava based cropping systems; (iv) to investigate changes in soil chemical and physical properties, generated by cropping systems and erosion over time.

Part 2 *MATERIALS AND METHODS*

The national territory of the Republic of Colombia can be sub-divided into five eco-regions, the hot and humid tropical lowlands of the Amazonian watershed (Amazonia) with 336,000 km², the hot and humid eastern grasslands draining into the Orinoco River (Orinoquía) with 266,300 km², the northern hot lowlands of the Caribbean (Llanura del Caribe) with 127,700 km², the hot per-humid forest areas of the Pacific Coast (Llanura del Pacifico) with 59,500 km² and the warm to temperate Andean Zone with 352,150 km². Coming from the South, the main string of mountains (Cordillera) starts to divide up into three mountain ranges, the Western, the Central and the Eastern Cordillera. Between the Cordilleras, two large river systems, the Cauca river and the Magdalena river develop, running from South to North and draining into the Caribbean. The Cauca valley being one of the Inter-Andean high valleys at approximately 1,000 m.a.s.l starts about half-way between the departmental capitals Popayan (capital of Cauca Department) and Cali (capital of Valle del Cauca Department), runs about 250 km in Northern direction and covers approximately 8,160 km². On the Andean slopes, in the transition zone between the Southern extreme of the Cauca valley and the Popayan highlands South of the valley, population and hence agricultural activities have increased dramatically in recent years. However, soil conservation measures have mostly not been adopted. The region is, therefore, a show-case for erosion and soil degradation processes and it provides perfect conditions for soil erosion research and soil conservation technology development.

2.1 *Locations and treatments*

The field trials, on which data are reported here, were conducted over two cropping periods, from May 1990 to April 1991 and from April 1991 to April 1992 under natural rainfall at two locations in the northern part of Department of Cauca, Colombia: *Santander de Quilichao* (3°6' N, 76°31' W, hereafter termed "Quilichao") and *Mondomo* (2°53' N, 76°35' W). The trial plots were established in 1986 at Quilichao and in 1987 at Mondomo and cropped with cassava or kept as continuously clean tilled fallow since.

The study sites are characterized by a subhumid climate. Rainfall is bimodally distributed, with maxima in May and October. Quilichao is located at the southern border of the Cauca valley with a gently ondulating topography. The trial was located at 990 m a.m.s.l. on the CIAT (Centro Internacional de Agricultura Tropical) research station. Annual mean temperature is 23.7° C and the average annual precipitation is 1,799 mm. The well drained soil has been classified as an amorphous, isohyperthermic Oxic Dystropept (Reining, 1992), developed from fluvially translocated volcanic, partly weathered material (IGAC, 1976). Mondomo, 20 km south of Quilichao, belongs to the Central Cordillera with moderate to steep, mainly convex slopes. The trial was conducted on a private farm at 1,450 m a.m.s.l. The nearest meteorological station is located 3 km away from the farm at 1,300 m a.m.s.l. with a mean annual temperatur of 21.5° C and a precipitation of 2,168 mm (15 year's average). The soil is a kaolinitic-amorphous, isohyperthermic Oxic Humitropept with excellent drainage (Reining, 1992). Parent material is an acidic magmatic material with a high proportion of volcanic ashes (IGAC, 1976). Some chemical and physical properties of topsoils in Quilichao and Mondomo are shown in part 2.5.

The trials were laid out on uniform slopes with gradients ranging from 7 - 13 % in Quilichao and from 12 - 20 % in Mondomo. The cropped plots were 16 m long and 8 m wide, separated by small grass strips. The bare fallow treatment consisted of plots with a length of 22.1 m and a width of 11 m. Small earth ridges, planted with grasses on the two sides and the upper ends of the plots were made to avoid soil and water losses from the plots and water inflow and sedimentation from outside. At the lower end small channels of asbestos cement were established to collect eroded soil and runoff.

The following treatments were laid out in three replications at Quilichao and two replications at Mondomo in a randomized incomplete block design:

(1) Bare fallow; continuously clean tilled.

(2) Planting of cassava (*Manihot esculenta* Crantz) on parallel contour ridges at 1 m distance as a sole crop in a 1 m x 1 m planting pattern.

(3) Planting of cassava on the flat as a sole crop at 1 m x 1 m.

(4) (only in Quilichao).
Planting of cassava on the flat at 1 m x 1 m, intercropped with two contour rows of *Pueraria phaseoloides* (Roxb.) Benth. (CIAT No. 9900) between the cassava rows at a distance of 0.25 m from cassava and 0.5 m between legume rows at a seeding rate of 4 kg ha^{-1}. The legume was seeded 2.5 months before cassava planting.
In the second cropping period cassava was planted in the existing legume sward; planting spots, "cajuelas", were manually prepared with a shovel.

(5) As in treatment No. 4, but intercropped with *Zornia glabra* Desv. (CIAT No. 8283); simultaneously intercropped at a seeding rate of 4 kg ha^{-1}.

(6) As in treatment No 5, but intercropped with *Centrosema acutifolium* Benth. (CIAT No. 5568) at a seeding rate of 3 kg ha^{-1}.

(7) (only in Mondomo).
As in treatment No. 5, but intercropped with *Centrosema macrocarpum* Benth. (CIAT No.5740) at a seeding rate of 3 kg ha^{-1}.

(8) Planting cassava on the flat at 1 m x 0.9 m with two contour barriers of *Vetiveria zizanioides* (L.) Nash. per plot at a distance of 8 m, occupying 12.5 % of the plot area. Each barrier consisted of two rows. Planting density of cassava was increased to compensate for the loss of effective cassava growing area (87.5 %).

(9) Planting cassava on the flat at 1 m x 0.8 m. Two contour strips per plot of *Pennisetum purpureum* Schumach. of 2 m width at a distance of 8 m were established, occupying 25 % of the plot area. Each strip consisted of three rows. Planting density of cassava was increased to compensate for the loss of effective cassava growing area. In Quilichao, the elephant grass barriers were already established in May 1989.

With the exception of treatment No. 4 legumes were established simultaneously as previous research done by CIAT (1979) showed stands of legumes to be poor, when planting was delayed 60, 210 and 300 days after cassava planting.
Table 1 shows planting and harvesting dates of cassava and cultivars used.

Table 1. Cultivars of cassava, planting and harvesting dates at Quilichao and Mondomo in the cropping periods 1990-91 and 1991-92.

Site	Cropping period	Planting date	Harvesting date	Cassava cultivar
Quilichao	1990-91	10 May	6 April	CM 507-37
	1991-92	22 April	5 April	CM 523-07
Mondomo	1990-91	18 May	18 January	"Seleccion40" [a]
	1991-92	11 April	21 January	MCol 2261

[a] local variety

Bare plots were treated according to instructions outlined by Wischmeier and Smith (1978). They were plowed up and down slope at planting time and kept free of vegetation by hand weeding. After every heavy erosion event plots were cultivated slightly to prevent severe crusting and gully formation. Soil preparation was done by a rotovator in Quilichao and by oxen plow in Mondomo, according to local tillage practices. Soils were tilled immediately

before cassava planting, except in the first cropping period in Quilichao, where tillage was done already three months before. After tillage, ridges were prepared manually in treatment No. 2. In each cropping period, dolomitic lime (54 % $CaCO_3$ and 46 % $MgCO_3$) was applied at a rate of 500 kg ha^{-1} and incorporated by tillage in cropped plots. Three weeks after cassava planting, 500 kg·ha^{-1} of a compound NPK fertilizer (10 % N, 20 % P_2O_5, 20 % K_2O) was applied around the cuttings. Weeding was done by hoe: in Quilichao twice in the first and three times in the second cropping period, in Mondomo three and four times, respectively.
Reseeding and irrigation for establishment of the legumes was necessary, especially for the small seeded *Z.glabra*. Legumes were cut regularly depending on their growth habit and vegetative characteristics. Plant material was taken off the plot.
Systemic insecticides were applied to control thrips and spider mites at Quilichao and thrips at Mondomo.
In both years cassava roots harvested at Mondomo were strongly attacked by white grubs (Phyllophaga sp.), which made an early harvest necessary after only eight (1991) and nine months (1992).

All plots except those of treatment No. 8 (cassava + vetiver grass barriers) were established in 1986 at Quilichao on an old grassland soil and 1987 at Mondomo on bush fallow. They were continuously cropped with cassava or under continuously clean tilled fallow (treatment No. 1). Reining (1992) reported results on soil erosion and productivity of cassava cropping systems, conducted over two cropping periods from 1987 to 1989 on the same plots. He put emphasis on different tillage systems and intercropping of cassava with grain legumes. In the following copping period different cassava cropping systems, including the application of mulch, were tested (CIAT, 1991a).

2.2 *Soil and runoff collection*

Sediments and runoff were collected in the channels established at the lower end of the plots. They were inclined slightly; sediments remained in the channel, whereas water flowed through a double filter into a sedimentation tank with a splitter at the upper side with 15 outlets at the same height level. The central outlet (6.7 % of total runoff) was connected to a barrel of 200 l storage capacity in the case of the cassava cropping systems.

This allowed the collection of a total runoff volume up to 23.4 l m^{-2}. Two barrels were established to catch the runoff from the bare fallow plots (up to 24.7 l m^{-2}).
Wet sediments were weighed and dry soil losses were calculated by determining the water content of representative samples. Remaining sediments in the runoff were measured and values added to the eroded soil which had been collected in the channels. Runoff was corrected for water remaining in the sediments and direct rainfall into the collection channel. In heavy rainstorms, some runoff overflow was observed, so results should be taken as minimum estimates. Measurements of runoff began in July 1990 in Quilichao and in September 1990 in Mondomo after restoration of the runoff infrastructure.
Soil loss measurements began in April 1990 and ended in April 1992. Sediments from bare fallow plots were collected after every erosive rain. On very few occassions, two or more erosive events occurred over night. In this case individual rainfall quantities were added up to match the single soil loss figure obtained for that night. At the beginning of the cropping period, soil and water losses from cropped plots were measured after every erosive rainfall and later on sediments monthly. Runoff was measured, whenever it occurred.

2.3 *Measurement of soil cover, cassava plant height, and forage and cassava yields*

Soil cover and plant height of cassava were determined biweekly by measuring two diameters of each of ten plants per plot. Based on their average, ground cover of cassava was calculated as a circle (Reining, 1992) and corrected by the percentage of soil cover within the circle, determined by a sighting frame (Stocking, 1988). Cover by legumes, weeds and mulch was estimated visually at the same dates as cassava.
Fresh root and above ground biomass yield of cassava were determined. After every cutting forage legumes were weighed and dry matter production was calculated after oven drying.

2.4 *Physical soil properties*

The *particle size distribution* of the permanent bare fallow soils was determined with the Pipette method (Day, 1965) at the "Instituto Geografico Augustin Codazzi" (IGAC) in Bogotá, Colombia.
The Hydrometer method (Gee and Bauder, 1986) was used to determine sand, silt and clay contents of bare and cropped soils and sediments. Organic matter was not destroyed

previously. Sodium-hexametaphosphate was used as a dispersing agent. These analysis were conducted at the CIAT soil laboratory in Palmira, Colombia.

The *stability of aggregates* of topsoils (0 - 10 cm) was determined by wet-sieving with a modified YODER method (Franco and Gonzalez, 1967) at the soil laboratory of the University of Palmira, Colombia: 25 g of air dry aggregates with diameters of about 10 mm, sampled in the field between field capacity and permanent wilting point (Kemper and Rosenau, 1986), are prewetted for 15 minutes, followed by the wet-sieving procedure. Stability was determined for the size ranges 2-10 mm, 1-2 mm, 0.5-1 mm, 0.25-0.5 mm and < 0.25 mm. The proportion of primary particles was very low and had not to be subtracted. Meanweight diameters were calculated as the sum of the products of the mean diameter of each size fraction and the proportion of the total sample weight, occurring in the corresponding size fraction (Van Bavel, 1949). Measurements were repeated twice for each plot in February 1991 and March 1992.

The *dry aggregate size distribution* of surface soils (0 - 10 cm) and of rain sorted aggregates, overlying soil crusts of bare fallow soils after heavy rain events, were determined by sieving with a set of sieves with diameters of 10, 5, 2 and 1 mm, respectively.

The *water permeability of soils* was determined by the double-ring infiltrometer method. It consisted of two concentric metal rings of 50 and 20 cm diameter. They were carefully driven into the soil to a depth of 10 cm and filled with water. The water intake was measured in the inner ring in time per centimeter of water height until a constant value was reached, generally after three to six hours. Practical reasons did not allow the use of wider rings, as suggested by Bouwer (1986). Two infiltration runs were made per plot in 1991 and 1992 in the dry season.

Undisturbed soil cores (97 cm^3, 4 cm high) were used to determine *bulk density* and *moisture retention*. Water contents at potentials of -1, -6, -10, -30, -100 and -1500 kPa were determined at the end of the first cropping period, using a pressure plate extractor (Richards, 1965). Six cores were taken from topsoils (0-20 cm) of each replication of bare fallow and cropped treatments. Measurements were conducted at the soil laboratory of the CENTRO NACIONAL DE INVESTIGACIONES DE CAÑA (CENICAÑA) near Cali, Colombia.

2.5 *Chemical soil and water properties*

Soils, sediments and water runoff were analysed at the CIAT soil laboratory near Palmira, Colombia. Representative samples were taken from the sediments. Chemical properties of topsoils at soil depths of 0-10 cm and 10-20 cm were monitored by analysing composite soil samples of each plot before the beginning of the cropping seasons (May 1990 and April 1991) and after harvesting of cassava in April 1992. Soils were sampled before the application of lime and fertilizer.

Water in the collection tanks was stirred thoroughly and one or, when the amount of runoff was high, two aliquots of 1 l were sampled. Water of events with a runoff amount of less than 1 mm m^{-2} was not analysed, because collection channels were exposed to direct rainfall and dilution effect was thought to be too large to give reasonable results. Sediments and runoff water were obviously contaminated by calcium possibly originating from the asbestos cement of the collection channels. Therefore, calcium could only be determined for sediments from the bare fallow plots, when enough soil was lost, to take uncontaminated samples. Soluble nutrients analysed in runoff were phosphorus, potassium and magnesium.

The *pH* of the soil was measured in a watery suspension of soil:H_2O = 1:1 with a glass electrode (Salinas and Garcia, 1985). *Organic matter* was analysed according to the method of Walkley and Black (Salinas and Garcia, 1985) by wet ashing with a $K_2Cr_2O_7$-H_2SO_4-solution and photometric measurement of the Cr^{3+}-content. *Total nitrogen* was determined with the method of Kjeldahl (Bremner and Mulvaney, 1982). *Mineral N* was extracted from soil with KCl and quantified directly by colorimetric methods (Keeney and Nelson, 1982). *Potassium* was extracted with a NH_4Cl-solution (Salinas and Garcia, 1985), *magnesium, calcium and aluminium* with a KCl-solution which was titrated with NaOH (Salinas and Garcia, 1985). *Available phosphorus* was analysed with a modified Bray-II method (Salinas and Garcia, 1985). After fusion with Na_2CO_3 *total P* was determined colorimetrically in the water soluble extract (Olson and Sommers, 1982). *Organic P* was analysed with the extraction method (HCl and NaOH) as the difference between total P and inorganic P (Metha et al., 1954; Olson and Sommers, 1982). *Sulphur* was extracted by a $Ca(H_2PO_4).H_2O$ solution and analysed colorimetrically (Salinas and Garcia, 1985). *Zinc, copper, iron,* and *manganese* were extracted with HCl + H_2SO_4 (Salinas and Garcia, 1985) and *boron* was determined by spectrophotometer in a solution buffered with CH_3COONH_4 + CH_3COOH and NaCl (Salinas and Garcia, 1985).

Table 2. Some chemical and physical properties of top soils (0-20 cm) and infiltration rates of permanent bare fallow, cropped and uncultivated soils at Quilichao and Mondomo in May 1990 [a].

	QUILICHAO			MONDOMO		
		---cropped---			---cropped---	
	bare fallow	since 1986	grass -land	bare fallow	since 1987	bush [b] fallow
pH (1:1)	4.1	4.5	4.6	4.3	4.6	4.7
OM (%)	5.3	6.0	7.3	4.4	6.3	7.1
total N (mg kg^{-1})	1923	2218	2072	1708	2053	2352
mineral N (mg kg^{-1})	56	51	51	63	67	70
P-BrayII (mg kg^{-1})	3.9	14.3	1.4	1.6	9.3	3.9
K (cmol kg^{-1})	0.08	0.17	0.16	0.08	0.22	0.29
Ca (cmol kg^{-1})	0.20	1.44	0.85	0.28	1.35	1.10
Mg (cmol kg^{-1})	0.05	0.54	0.30	0.06	0.48	0.56
Al (cmol kg^{-1})	4.78	3.24	4.03	2.25	2.13	2.90
S (cmol kg^{-1})	81.3	64.0	39.3	63.5	63.3	65.0
B (cmol kg^{-1})	0.24	0.30	0.22	0.15	0.29	0.37
Cu (cmol kg^{-1})	0.71	0.77	0.72	0.96	1.04	0.49
Zn (cmol kg^{-1})	0.52	0.76	0.59	0.57	0.67	1.40
Mn (cmol kg^{-1})	2.1	9.2	8.8	88.6	55.0	25.1
Fe (cmol kg^{-1})	36.7	30.1	37.9	26.2	33.4	18.6
Sand (%)	23.3	20.2	24.6	24.9	24.2	35.6
Silt (%)	17.5	19.3	19.2	19.0	20.3	26.0
Clay (%)	59.2	60.5	56.2	56.1	55.5	38.4
infiltration rate [c]	5.9	3.5	4.8	4.9	12.7	8.3
bulk density (g cm^{-3})	1.02	1.03	1.06	1.02	1.01	0.90
degree of aggregation [d]	77.3	79.5	93.0	60.0	66.9	85.7

[a] Values are means of three replications in Quilichao and two in Mondomo and of two depths per plot (0-10 cm and 10-20 cm).
[b] Only one replication was considered for bush fallow soils in Mondomo. The second one was used as a home garden before, which received chicken manure. Its soil properties were not representative for the typical cassava soils in the region.
[c] Final rates in $cm \cdot h^{-1}$
[d] Percentage of particles > 0.5 mm, determined with a modified Yoder method (Franco and Gonzalez, 1967).

2.6 *Precipitation characteristics*

Monthly recording raingauges (according to Hellmann, Model 1509-20, Lamprecht, Göttingen, Germany) were established at both experimental sites adjacent to the erosion plots to measure rainfall amounts, calculate intensities and erosivity (R) factors of the Universal Soil Loss Equation (USLE; Wischmeier and Smith, 1978) as well as several additional erosivity indices.

2.7 *Calculation of erosivity indices*

Coefficients of correlations for the following erosivity indices with soil (and water) losses from continuously clean tilled fallow plots were tested:

R-factor of the *USLE*:
In the eastern part of the USA the storm to storm variation in soil loss was highly correlated to the product of the total rainfall energy and the maximum 30-minute intensity of a storm ($E_u \cdot I_{30}$), the so called R-factor of the USLE.
The R-factor is calculated according to the procedures used for the USLE (Wischmeier and Smith, 1978): rain events less than 12.7 mm and separated from other rain periods by more than six hours were omitted, unless at least 6.35 mm fell in 15 minutes. Rainfall is subdivided into ranges of equal intensities (I) and kinetic energy (E_u) per mm of precipitation is calculated as follows:

$$\text{for } I <= 76 \text{ mm h}^{-1}: \quad E_u \text{ (USLE)} = 0.119 + 0.0873 \log_{10} I$$

$$\text{for } I > 76 \text{ mm h}^{-1}: \quad E_u \text{ (USLE)} = 0.283$$

where I has units of mm h^{-1}, E_u of megajoules per hectare per millimeter of rainfall (MJ ha^{-1} mm^{-1}).
The individual R-factor of a rain event is the accumulated kinetic energy of all increments multiplied by its maximum intensity sustained for 30 minutes ($E_u \cdot I_{30}$). The sum of all individual storm values during one year is the annual USLE R-factor.
Soil losses of all plots were adjusted by the topographic factor (LS) of the USLE to that of 'unit plot' conditions (22.1 m length and 9% slope), if not otherwise indicated.

Rainfall amount: A

A is the most simple index related to particle detachment and transport by water runoff.

Energy: E_u

E_u was calculated according the formula for kinetic energy calculations of the USLE. The storm energy is an indication for the volume of rainfall and for potential runoff.

Energy: KE

On one research station in Zimbabwe, Hudson (1971) observed that erosion occured only, if rain intensity exceeded about 25 mm h^{-1}. Energy was calculated as follows (I in mm h^{-1}):

$$KE{>}25 = 29.8-(127.5\ I^{-1}) \quad \text{(measured in J m}^{-2}\text{ mm}^{-1}\text{)}$$

Intensities:

Raindrop erosion increases with rain intensity. I_{max} indicates the prolonged peak rates of detachment and runoff (Wischmeier and Smith, 1978). I_{max} is determined for maximum intensity sustained during 5, 7.5, 10, 15, 20, 30, 45, or 60 minutes. The *square* of rain intensity is strongly related to interrill erosion, as outlined by Foster (1982).

Products of all single parameters:

$A\ I_{max}$ $E_u\ I_{max}$ $E_u\ I_{max}\ A$ $E_u\ A$

The products of rain amount or kinetic energy (USLE) and maximum intensity indicate how particle detachment is combined with transport capacity. The best suited duration of the peak intensity of a rain event is searched for the conditions of the study sites.

Rainfall amount multiplied by intensity including a term for the *soil moisture conditions prior to erosive rain events:*

$$(A + (An \cdot n^{-1})) \cdot I_{max}$$

for accumulated rainfall in n = 5, 4, 3, 2, 1 days before the erosive event. Antecedent rainfall amount was then divided by the number of the days included in the calculation to account for water losses by evaporation and water infiltration (modified from Stocking and Elwell, 1973).

EIA:

EIA is the product of I_{30} multiplied by the square root of the product of rainfall and runoff volume (Foster et al., 1982). In this case I_{30} is effective over the duration of the runoff phase, unlike the R-factor (E_u I_{30}) of the USLE, where I_{30} is assumed to be effective over the duration of the entire storm.

2.8 *Calculation of erodibility (K of the USLE)*

Direct measurements: On the bare fallow plots (USLE) factors for slope length (L), crop management (C) and support practices (P) are equal 1.0. Soil losses were corrected for the slope gradient factor (S) to be equivalent to those of standard bare fallow plots (USLE-"unit" plot: 22.1 m long and 9 % slope). Soil erodibility (K) then is the amount of soil loss (t ha^{-1}) divided by the corresponding erosivity factor R (MJ mm ha^{-1} h^{-1}; per storm event, month, year):

$$K = (\text{soil loss}) \cdot (E_u \cdot I_{30})^{-1}$$

Nomograph solution: Based on the work of Wischmeier and Mannering (1969) on the relationship between soil properties and erodibility, Wischmeier et al. (1971) developed a nomograph to predict the susceptibility to erosion from a few easily measured soil parameters, including a term for texture, soil structure, permeability and the percentage of organic matter. K-values are calculated according to the following equation (Wischmeier and Smith,1978; Foster et al., 1981):

$$K = 0.277 \cdot (10^{-6}) \cdot M^{1.14} \cdot (12-o) + 0.0043 \cdot (s-2) + 0.0033 \cdot (p-3)$$

where M is the percentage of silt + very fine sand (0.1 - 0.002 mm) multiplied by the sum of percentages of silt and sand; o is the percentage of organic matter, s the soil-structure code and p the profil-permeability class, used in soil classification (USDA, 1951).

2.9 *Crop management factors (C of the USLE)*

Crop management factors (C) of the USLE are defined as the ratio of soil loss from land cropped under specific conditions to the corresponding loss from clean tilled, continuous fallow (Wischmeier and Smith, 1978). *Preliminarly* crop stage periods are defined as follows (Wischmeier, 1960; modified):

Period 1 - tillage for seed bed preparation to two month old cassava
Period 2 - two to five month old cassava
Period 3 - five to ten month old cassava
Period 4 - harvest to tillage for the following crop.

2.10 *Data analysis*

Data of all treatments and both locations were analysed using general linear models procedures (GLM) with SAS (1988). Yearly data were regarded as repeated measurements and analysed by the difference between years in the same models as for the individual years. Residuals of data, that were not obviously normally distributed, were tested by the Shapiro-Wilk test. (If necessary, square root transformation was used to achieve normality). Treatment means were compared to the control (bare fallow) with the Dunnet Test, and among cropping systems with the Tukey Test. The year effect (difference) for each treatment was tested by the T-Test for H0 = 0. Correlations between variables were analyzed separately for each location for the treatment means of replications. Standard linear regression procedure was used to determine the equation for soil loss and erosivity parameters.

Figure 1: Traditional oxen ploughing for cassava planting at Quilichao.

Figure 2: Cassava cropping on heavily eroded hillsides in the tropical Andes of South-west Colombia.

Figure 3: A continuously clean tilled *Wischmeier and Smith* standard plot at Quilichao. High amounts of lost soil can be appreciated in the background at the bottom of the plot.

Figure 4: View of the erosion trial in Mondomo with a self-registering rain gauge in the foreground recording rainfall amount over time from which intensity is calculated.

Figure 5: Determination of infiltration rates with the double-ring infiltrometer method in the cassava-*Penisetum purpureum* treatment.

Figure 6: Although providing soil cover later, sole cropped cassava during early growth stages does not provide much protection from erosive rains.

Figure 7: Cassava-*Zornia glabra* intercropping two months after sowing at Quilichao. Due to poor legume establishment little soil cover developed initially allowing heavy erosion losses on the bottom part of the plot.

Figure 8: Cassava-*Pueraria phaseoloides* intercropping in the second cropping period at Quilichao. Soil cover was much better but strong competition with cassava can be noted.

Figure 9: Soil cover by cassava-*Stylosanthes guyanensis* intercropping in the dry season was excellent but competitive effects lead to reduced vigour and canopy development in cassava.

Figure 10: Grass barriers planted to *Pennisetum purpureum* recovered quickly after each cut-back but agressive, horizontally expanding root systems reduced cassava growth and yield in adjacent rows.

Figure 11: Cassava cropping with *Vetiveria zizanioides* grass barriers. This grass forms a gapless barrier, effectively controlling erosion whilst competing little with cassava.

Figure 12: *Arachis pintoi* was used to close gaps in newly established vetiver grass contour barriers.

Figure 13: Traditional cassava cropping on eroded hillsides in the study region near Mondomo, photographed in 1983, three years before the project started to operate.

Figure 14: Farmers trying out different grass species as barriers and barrier distances to find their own solution to the erosion problem. Picture taken in 1993, three years after project had started its on-farm phase.

Part 3 *EROSIVITY AND ERODIBILITY*

Soil loss of agricultural land by water erosion is a function of precipitation characteristics, soil properties, topography, crop and soil mannagement. The best known and most widely used prediction equation is the Universal Soil Loss Equation (USLE; Wischmeier and Smith, 1978); it has been successfully applied to conditions similar to those, where it was developed. Extensive research was done in temperate and subtropical areas to estimate erosion potential, to design effective erosion control measures and to use it as a tool for adequate soil conserving land use planning. However in the tropics, where the erosion hazard is more serious, the applicability of the USLE is questionable. Heavy rainfall of high intensity is much more frequent. Soil parent material, rate of soil formation and susceptibility to erosion can differ considerably from that under temperate conditions. Also, more and more steep slopes are cropped with differentcrops and cropping practices. For these reasons there is an urgent need to test and to determine the factors of the USLE and if necessary look for an appropriate modification. This also applies to the tropical Andes where information on the susceptibility of soils to erosion and the potential effects of rain are particularly scarce.

The product of the kinetic energy of a rain and its maximum intensity sustained for 30 minutes, the so called R- or Erosivity factor of the USLE, explained best ($r^2 = 0.72 - 0.97$) the variation in soil loss in the eastern USA under different conditions of climate, soil and topography (Wischmeier, 1959). This combined energy-intensity term accounts for the detachment and transport of soil particcles by raindrop impact and overland flow. Under some tropical and sub-tropical conditions the R-factor proved to be correlated quite satisfactorily with soil loss on bare fallow plots: in Zimbabwe on two research stations (Stocking and Elwell, 1973), in Kenya (Ulsaker and Onstad, 1984), in Hawaii (Lo et al., 1985), in the Southern Guinea Savanna of Western Nigeria (Sabel-Koschella, 1988), in Indonesia (Adimihardja, 1989), in Venezuela (Paez and Rodriguez, 1989). Better explanations of soil loss variation was found by Hudson (1971) on a research station in Zimbabwe with the kinetic energy of that part of the rain, falling with intensities over 25 mm h^{-1} (KE > 25). In experiments conducted in Western Nigeria (Lal, 1976a), Trinidad (Gumbs et al.,1985) and Sierra Leone (Bomah, 1988), the product of rainfall amount and its maximum intensity in 7.5 minutes correlated best. Ulsaker and Onstad (1984) replaced the energy term of the R-factor with the amount of rainfall and got somewhat higher correlations than with the R-factor. On the Caribbean island of Tobago

low correlations of soil loss with the R-factor were found, but they increased somewhat, when I_{30} was replaced by I_{15}, the maximum intensity during 15 minutes of a rainstorm (Ahmad and Breckner, 1974).
Stocking and Elwell (1973) included a simple index of rainfall prior to the erosion event to account for the influence of the antecedent moisture conditions on erosion, which improved the correlations with soil loss only marginally.
Generally the susceptibility of tropical soils to erosion is considered to be lower than that of soils developed in temperate climates. Soil Erodibility values (SI-units) are reported to range from 0.004 to 0.09 for 23 major soils in the United States (Wischmeier and Smith, 1978). Dangler and El-Swaify (1976) reported a wide range of K-values for Hawaiian soils, similar to that found for temperate soils in the USA. Even within a single soil order, the range of variation was as wide as it was between orders. The order Inceptisols included both very erodible and very erosion resistant soils. Barnett et al. (1971) found low erodibility values for two Inceptisols in Puerto Rico presumably due to their high organic matter content and excellent structural properties. Adimihardja (1989) reported moderate K-values for two Inceptisols in Indonesia; they were higher than those of Oxisols and Ultisols. Erodibility values computed by the USLE nomograph and measured directly compared reasonably well for nine non gravelly soils studied in the Ivory Coast (Roose, 1977) and in Indonesia (Kurnia et al., 1986 cited by Adimihardja, 1989); however serious discrepancies were apparent for selected Hawaiian soils with amorphous and oxidic constituents (El-Swaify, 1977). Studies on well aggregated soils in the USA show, that nomograph predictions significantly underestimated their K-values (Röhmkens et al., 1975; Young and Mutchler, 1977).

3.1 *Rainfall erosivity in the tropical Andes*

3.1.1 Rainfall characteristics

Amount, intensity and kinetic energy of rainfall, are generally much higher in the humid and subhumid tropics than in temperate regions. The high erosion potential in the tropics is attributed more to high rain erosivity than soil erodibility. High intensities exceed infiltration rates easily, even on very permeable soils such as those found on the study sites, causing runoff and soil loss. Soil aggregates are broken up by raindrop impact, particles block soil pores or are moved downward by splash and runoff water.

Table 3. Characteristics of rainfall and USLE R-factors in Quilichao and Mondomo.

	QUILICHAO				MONDOMO			
	1987-1988 [a]	1988-1989 [a]	1990-1991	1991-1992	1987-1988 [a]	1988-1989 [a]	1990-1991	1991-1992
Rainfall								
total (mm)	1458	2417	1529.1	1654	1498	2377	1261	1274.5
% erosive	80.2	77.8	63.9	65.4	71.2	73.9	68.6	45.9
No.erosive events	--	--	37	31	--	--	32	23
No. USLE - events [b]	39	58	49	44	41	55	38	31
Intensity distribution (% of rain amount)								
> 25 mm h^{-1}	40	40	40	40	34	36	49	30
> 50 mm h^{-1}	21	18	17	20	17	18	24	15
> 75 mm h^{-1}	11	8	15	9	5	8	12	10
Kinetic energy								
total (MJ ha^{-1} a^{-1})	310.8	512.2	324.7	359.1	304.1	488.1	273.2	255.6
% erosive	--	--	68.7	69.6	--	--	75.2	51.9
J m^{-2} mm^{-1} total	21.3	21.3	21.2	21.7	20.3	20.5	21.7 [c]	20.1
J m^{-2} mm^{-1} erosive	--	--	23.1	23.0	--	--	23.9	22.7
R-factor								
(MJ ha^{-1} mm h^{-1} a^{-1})[d]			9042	11078			7548	5142

[a] Data from Reining (1992).
[b] USLE events are rains, which fulfill the conditions, outlined by Wischmeier and Smith (1978) for R-factor calculations.
[c] 21.5 J m^{-2} mm^{-1} for 80 % of the rainfall. For reasons of reparation of the raingauge in 1990-91 in Mondomo, four heavy rainstorms had to be excluded. Their kinetic energies and R-factors were estimated.
[d] R-factors are given on a "per cropping season" basis with durations of 341 and 373 days in Quilichao, and durations of 341 and 379 days in Mondomo for the 1990-91 and 1991-92 seasons, respectively.

Rainfall in Quilichao was somewhat lower than average, 1529 mm in the first, 1654 mm in the second study period, whereas in Mondomo it fell well below the long term average with 1261 and 1275 mm (Table 3). Fig. 15 shows the percentage of rain falling in different size classes: The higher contribution of events up to 25 mm in Mondomo with 64.4 % of total rainfall is remarkable, compared to 46.6 % in Quilichao, however, rainfall events between 25 and 50 mm accounted for 32.1 % in Quilichao and 20.3 % in Mondomo. Hudson (1971) states that in temperate climates 5 % of rainfall exceeds a threshold intensity of 25 mm h^{-1}, above which he considers rainfall to be erosive, while in the tropics 40 per cent of rain is falling with a higher intensity. This compares well with the rainfall in Quilichao; in Mondomo nearly 50 % exceeded this threshold value in the first, and 30 % in the second year (Table 3).

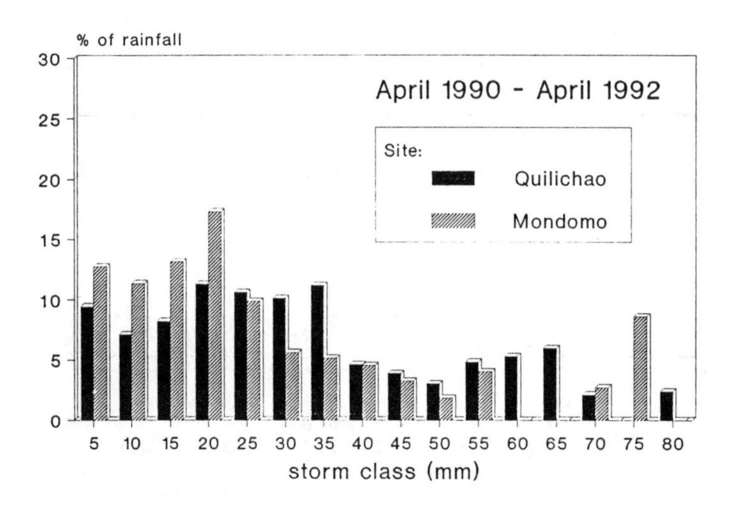

Figure 15: Percentage of rainfall in different storm size classes at Quilichao and Mondomo.

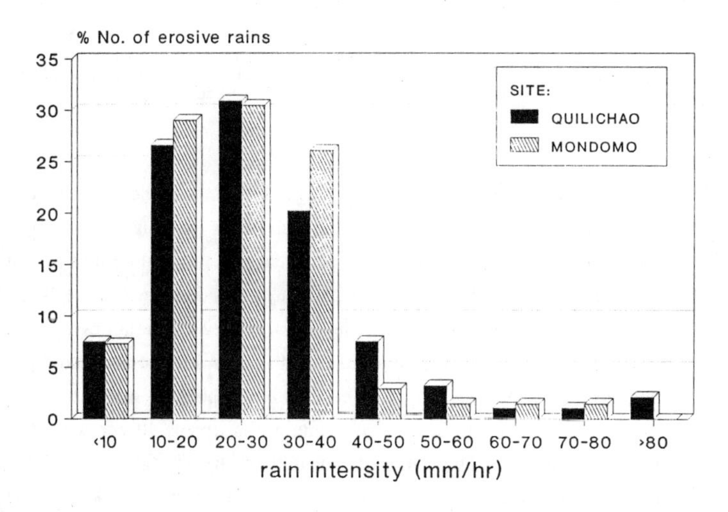

Figure 16: Distribution of the maximum rain intensities sustained for 30 minutes (I_{30}).

Maximum intensities measured in the two-year period were 330 mm h^{-1} in Quilichao, when 11 mm of rain fell in two minutes, and 180 mm h^{-1} in Mondomo, also sustained for two minutes. The maximum 30 minute intensity of the greater part (78 % in Quilichao and 86 % in Mondomo) of all USLE rain events which are included in the R-factor calculations ranged from 10 to 40 mm h^{-1} (Fig. 16).

The proportion of rain falling in different intensity classes is given in Fig. 17 for total rainfall and for the ten storms which produced the greatest soil losses. These events showed a very even distribution. Erosive rainfall showed a weak trend towards higher rates of precipitation, falling at the lower intensity ranges in Mondomo and a somewhat increased rate at the higher ranges in Quilichao.

Average energy load for the two-year period was 21.5 J m^{-2} mm^{-1} in Quilichao for total and 23.0 J m^{-2} mm^{-1} for erosive rainfall. In Mondomo, these values were 20.9 and 23.3, respectively (Table 3). Similar results were obtained by Reining (1992) for the study sites during 1987-88 and 1988-89: The energy of total rainfall was 21.3 J m^{-2} mm^{-1} in Quilichao and 20.4 J m^{-2} mm^{-1} in Mondomo. The kinetic energy term (USLE) is derived from intensity-kinetic energy relations of temperate rainfall (based on drop size data of Laws and Parsons, 1941), however, its application in the tropics may be doubtful. The unit rainfall energy as a function of intensity varies with location over the world (Hudson, 1971). Drop size measurements of rain in the study region are necessary to clarify the unit energy-intensity relationship. Computed USLE values of energy overestimated the average values for kinetic energy in Zimbabwe (Hudson, 1961) and for five raintypes (air mass, precold front, warm front, easterly wave and trough aloft) at Miami, USA, especially at high rainfall intensities (Kinnell, 1973). Kowal and Kassam (1976) measured the energy load of rainstorms in Northern Nigeria and related kinetic energy to the rainfall amount per storm. They found an average load for 18 rainstorms in Northern Nigeria of 36.7 J m^{-2} mm^{-1}, ranging from 21.8 to 39.9 J m^{-2} mm^{-1}. Elwell and Stocking (1973) reported an average load of 19.1 J m^{-2} mm^{-1} from one station in Zimbabwe.

Not all rain events, which fulfilled the conditions outlined by Wischmeier and Smith (1978) for $E_u \cdot I_{30}$ calculations caused erosion (Table 3): in Quilichao only 37 out of 49 USLE rain events in the first and 31 out of 44 in the second study period produced soil loss. Precipitation with an amount higher than 25 mm or an I_{30} exceeding 30 mm h^{-1} occurred without soil loss. This was the case when rain fell on recently tilled soil or under dry soil conditions. The same is valid for Mondomo. Here, in the second period, not all USLE rains caused soil loss, but additionally four rains less than 12.7 mm and an intensity of less than 6.35 mm in 15 minutes generated erosion (R = 109.2 MJ mm ha^{-1} h^{-1}).

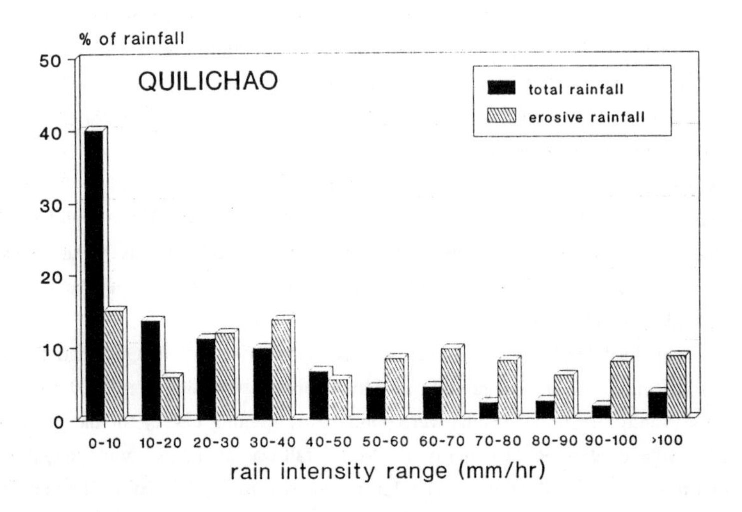

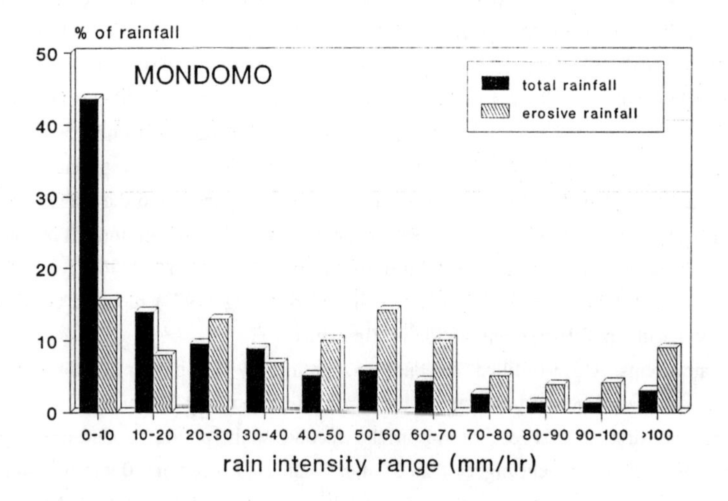

Figure 17: Proportion of rain falling at different intensity ranges for total rainfall and for the ten storms producing greatest soil losses in Quilichao and Mondomo.

However, the contribution of these four events to the total annual soil loss with 0.53 t ha^{-1}, less than 1 %, was very low.

3.1.2 Losses of soil and water from continuously clean tilled fallow

USLE adjusted monthly soil losses and corresponding precipitation as well as USLE R-factors for the periods 1990-91 and 1991-92 are shown in Fig. 18a and 18b. Soil loss data (not corrected by the USLE S-factor) from the previous years (1987 to 1990) are presented in Table 6. The erosion potential is high: In only two years (1990-1992) about 3.4 cm of topsoil in Quilichao and 1.5 cm in Mondomo was lost by erosion (bulk density 1.02 g cm^{-3}). Higher losses in Quilichao can be explained by the higher rain erosivity and the higher soil erodibility. Farmers in the study region generally crop much steeper slopes than those used in this study, presenting a very high erosion risk. Furthermore, soils are already degraded to a large extent; soil cover development is slow and annual crops such as cassava, beans and maize are normally grown without erosion control practices. Higher losses in Quilichao in the second year (April 1991 - April 1992) were caused by three extraordinary rainstorms producing 60.5 % of the total annual soil loss and accounting for 30 % of the annual R-factor. The highest single erosion event yielded 44.2 t ha^{-1} (on one replication 57.7 t ha^{-1}) in Quilichao and 23.7 t ha^{-1} in Mondomo. The 10 to 20 most severe rain events determine the level of annual erosion losses, as stated by Roose (1977) for dry and humid tropical environments. These events were associated with severe rilling and produced over 90 % of the yearly losses in Quilichao and over 95 % in Mondomo.
Considering these high soil losses, similarly high amounts of runoff could be expected. The proportions of rain lost as runoff, however, were low ranging from 10 to 18.2 % when no correction for slope gradient was applied. At Quilichao up to 70 % and at Mondomo up to 53 % of rainfall ran off in single events.

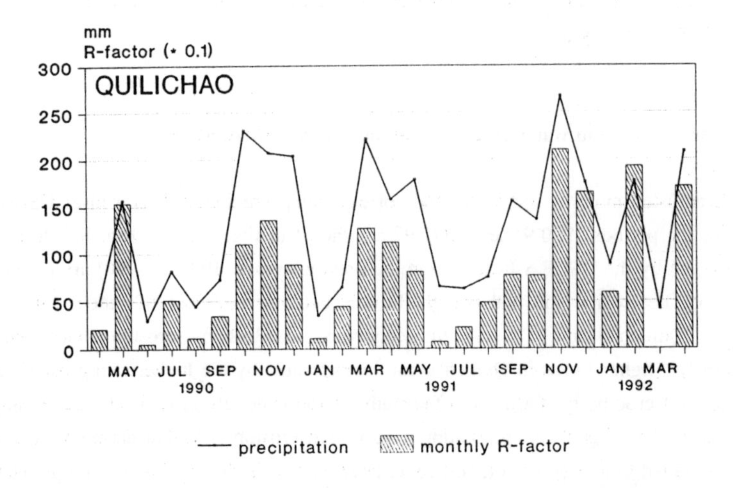

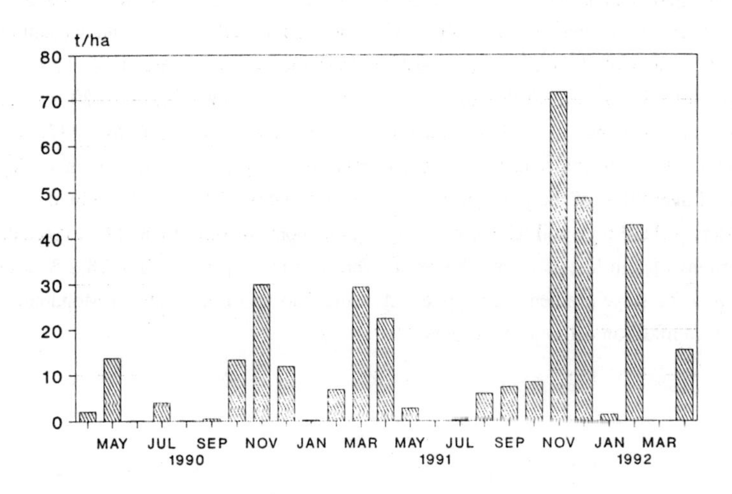

Figure 18a: Monthly amount and erosivity of precipitation (above) and corresponding dry soil loss (below) from permanent bare fallow plots (22.1m length and 9% slope; corrected by USLE-factors) in Quilichao

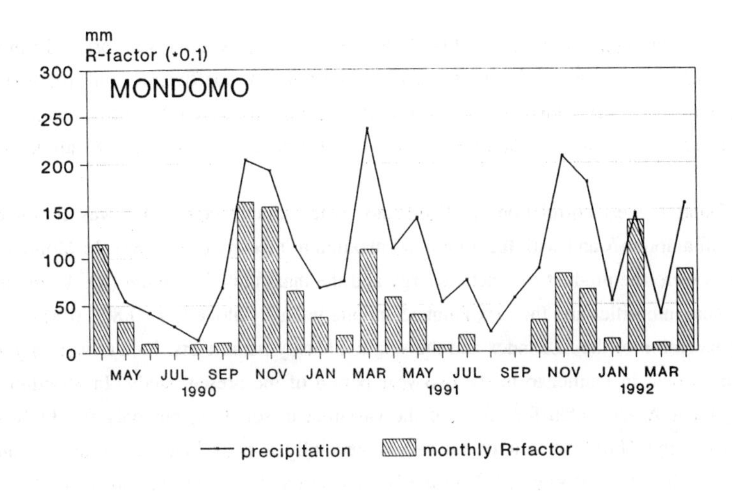

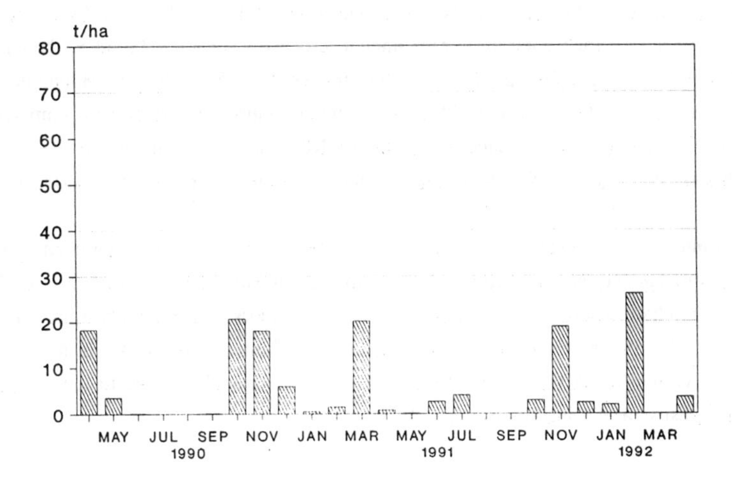

Figure 18b: Monthly amount and erosivity of precipitation (above) and corresponding dry soil loss (below) from permanent bare fallow plots (22.1m length and 9% slope; corrected by USLE-factors) Mondomo.

3.1.3 Erosivity indices

Simple correlation analysis was used to fit the erosivity factors to the soil loss and runoff data for each plot, season and year. Selected correlation coefficients and regression parameters for soil loss data over the two year's period, based on 57 measurements in Quilichao and 47 in Mondomo, and Reining's data (Reining,1992) from 1988-89 are shown in Tab. 4.

Reining found highest correlations in Quilichao, when kinetic energy E_u was combined with rainfall amount A and with the 30 minute maximum intensity ($r = 0.83$). In Mondomo best indices were the product of kinetic energy and the maximum 30 minute (R), 15 minute or 7.5 minute intensities and the maximum 15 minute intensity alone ($r = 0.80 - 0.81$).

With the exception of the *EIA* index and the single intensity parameters, indices were higher in Mondomo than in Quilichao in the two-year period of the present study. In Mondomo, rainfall amount A accounted for 56 % of the variation in soil loss, but only for 32 % in Quilichao. In the USA, amount of rainfall is generally a poor index (Wischmeier and Smith, 1958), however, Roose (1973) in the Ivory Coast and Ulsaker and Onstad (1984) in Kenya also found significant correlations. Data from the USA show, that the kinetic energy itself is not a good indicator of erosivity. A long slow rain may have the same E_u value as a shorter rain at much higher intensity (Wischmeier and Smith, 1978). In Quilichao the energy terms were clearly superior to rain amount, especially for KE>25 index, the energy of that portion of a rain falling with intensities greater than 25 mm h^{-1}. In Mondomo, the USLE kinetic energy term and KE>25 only marginally improved correlations compared to rain amount. Reining (1992), stated that Hudson's KE factor greatly improved correlations on both study sites during 1988-89 (Table 4), but they were lower than those found in this study.

The best intensity parameter was I_{30}; the coeffficients of correlation were improved by using the square of I_{30} and increased from $r = 0.76$ to 0.81 in Mondomo and from $r = 0.77$ to 0.85 in Quilichao, where I_{30} was among the erosivity factors with the highest accuracy of prediction. This probably indicates greater importance of interrill soil losses in the erosion process in Quilichao on flat slopes. This erosion type is strongly related to the square of rainfall intensity, as outlined by Foster (1982).

Table 4. Coefficients of correlation (r) and corresponding regression equation parameters (intercept; slope) for the relationship between soil loss from permanent bare fallow plots and selected erosivity indices in Quilichao and Mondomo from April 1990 to April 1992 [a] and correlation coefficients from April 1988 to April 1989.

	QUILICHAO				**MONDOMO**			
	1988 [b] -1989	1990 - 1992			1988 [b] -1989	1990 - 1992		
Indices of Erosivity [c]:	r	r	intercept	slope	r	r	intercept	slope
A	0.29^{ns}	0.57	-4.82	3.08	0.48^{*}	0.75	-2.99	2.13
E_u	0.40^{*}	0.64	-5.89	0.02	0.56^{*}	0.78	-2.80	0.87
KE>25	0.59^{**}	0.71	-4.22	0.02	0.75	0.76	-1.00	0.01
I_{15}	0.56^{**}	0.67	-9.93	3.55	0.80	0.62	-4.15	1.65
I_{30}	0.67	0.77	-11.30	5.20	0.79	0.76	-4.67	2.60
I_{45}		0.76	-9.92	6.00		0.72	-4.17	3.13
I_{60}		0.72	-8.58	6.49		0.69	-3.83	3.64
$A \cdot I_{7.5}$	0.68	0.75	-3.86	0.47	0.79	0.86	-1.58	0.28
$A \cdot I_{15}$	0.68	0.77	-3.97	0.59	0.79	0.88	-1.43	0.34
$A \cdot I_{30}$	0.71	0.80	-3.27	0.71	0.72	0.90	-0.88	0.41
$A \cdot I_{45}$		0.77	-2.57	0.81		0.87	-0.73	0.50
$A \cdot I_{60}$		0.73	-1.95	0.87		0.84	-0.61	0.58
$E_u \cdot I_{15}$	0.73	0.79	-3.61	0.24	0.80	0.88	-1.16	0.13
$E_u \cdot I_{30}$ (R-factor)	0.75	0.81	-2.93	0.29	0.80	0.89	-0.65	0.16
$E_u \cdot I_{45}$		0.79	-2.32	0.32		0.86	-0.47	0.19
$E_u \cdot I_{60}$		0.75	-1.78	0.35		0.84	-0.36	0.22
$E_u \cdot A \cdot I_{30}$	0.83	0.77	+0.21	0.04	0.75	0.93	+0.68	0.02
I_{30}^2		0.85	-2.92	0.67		0.81	-0.69	0.35
$A \cdot I_{30}^2$		0.86	-0.28	0.11		0.90	+0.56	0.06
$E_u \cdot I_{30}^2$		0.86	+0.02	0.04		0.89	+0.68	0.02
$(A+A_{5/5}) \cdot I_{30}$		0.83	-3.74	0.64		0.90	-1.70	0.42
$(A+A_{4/4}) \cdot I_{30}$		0.83	-3.57	0.62		0.89	-1.59	0.41
$(A+A_{3/3}) \cdot I_{30}$		0.84	-3.34	0.60		0.89	-1.56	0.41
EIA		0.95	-0.58	0.37		0.93	+0.33	0.21

[a] Based on 57 measurements in Quilichao and 47 in Mondomo. Indices are correlated at a significance level of $P<0.001$, if not otherwise indicated: ns = not significant, * and ** = significant at levels of $P<0.05$ and $P<0.01$, respectively.

[b] Data from Reining (1992); results are based on 31 measurements in Quilichao, 27 in Mondomo.

[c] Detailed description of erosivity indices is found in Part 2 "Materials and Methods". A = rainfall amount; E_u = kinetic energy (USLE); KE > 25 = kinetic energy (Hudson, 1971); I_n = maximum intensity sustained during n minutes; $AI_{7.5} = AI_m$ (Lal, 1976); $(A+An/n) I_{30}$ = influence of rainfall (n=days) prior to the erosive event; EIA = $(A \text{ Runoff volume})^{-0.5} I_{30}$ (Foster et al., 1982).

At both locations, 97 % of the variation in total kinetic energy of a rain was explained by rainfall amount. For this reason, the products of either rainfall or kinetic energy times the maximum intensities showed very similar correlations. The $A \cdot I_{30}$ has the advantage of easy calculation. Best factors were found when I_{30} was used. The product of the USLE R-factor and rainfall amount correlated best in Mondomo with r = 0.93. Only in Mondomo the R-factor fell into the range of soil loss predictions, given by Wischmeier (1959) in the USA with determination coefficients of $r^2 = 0.72$ to 0.97.

Castillo F. (1994) reported low correlations of the R-factor in Mondomo (r = 0.30; not significant) in his study on soil erodibility, conducted from February 1992 to January 1993 on plots of the present study. In Quilichao significant correlations were found (r = 0.65; $P < 0.01$), but they were lower than those found in the present study. Rainfall was well below average on both locations in this time period. These results stress the importance of long-term investigation on rain erosivity. Due to the cyclical pattern of rainfall in the tropics with an unpredictable duration of wet and dry cycles, a continuous long-term collection and evaluation of rainfall data is necessary to obtain reliable R-values for a certain region and to examine R as an erosivity index (Wischmeier, 1959).

The amount of antecedent rainfall five days prior to the erosion event ranged from 0 to 126.5 mm in Quilichao and 0 to 114.5 mm in Mondomo. Compared to the $A \cdot I_{30}$ parameter, the inclusion of a simple antecedent rainfall term to account for the influence of soil moisture conditions prior to the erosion event, improved the coefficients of determination in Quilichao from $r^2 = 0.64$ to $r^2 = 0.71$. In Mondomo no improvement could be achieved. One reason might be the higher water permeability of soils in Mondomo, but as suggested by Stocking and Elwell (1973) for erosion trials in Zimbabwe, the measure for antecedent rainfall is perhaps too crude and does not take into account the many factors influencing evapotranspiration (they worked with bare and cropped plots).

Runoff volume alone accounted for 69 % of soil loss variation in Quilichao and for 72 % in Mondomo (data not shown). *EIA* was by far the best erosivity index in Quilichao (r=0.95). But results have to be analysed with caution as on a few occasions runoff exceeded the capacity of storage tanks. Runoff beyond storage capacity occurred on three occasions. lacking data were calculated using the regression equations for the highly significant soil loss-runoff relation. In this way, an estimate of total unaccounted runoff was made reaching 61 mm for the two year observation period in Quilichao and 70 mm in Mondomo. This corresponded to 15.6 and 23.4 % of total measured runoff for that period, respectively.

Correlation coefficients were much higher in the dry season (Tab. 5). Perhaps the higher water acceptance rate of dry, in this case highly permeable soils compensates the high erosivity of tropical rainfall and quasi-temperate conditions are created. Notable is the superiority of the $E_u \cdot I_{30}^2$ or $A \cdot I_{30}^2$ index in Quilichao; in both seasons correlations with soil loss were greatly increased. Generally high correlations were found in the dry season in Mondomo.

Correlation coefficients were generally higher for water than for soil losses, except for the single intensity parameters from I_5 to I_{20} (results not shown). High correlation coefficients were found, when the maximum 45 minute intensity was included in Mondomo, instead of I_{30} for soil losses. In Quilichao, however, I_{45} did not improve the correlations with water losses.

Table 5. Coefficients of correlation (r) of selected erosivity parameters in dry and wet seasons (April 1990 - April 1992) in Quilichao and Mondomo.

Erosivity Index	**QUILICHAO** (n [a]=20) dry	(n=37) wet	**MONDOMO** (n=16) dry	(n=31) wet
A	0.62**	0.54***	0.92***	0.68***
I_{30}	0.77***	0.78***	0.86***	0.72***
E_u	0.68**	0.62***	0.93***	0.72***
$A \cdot I_{30}$	0.91***	0.74***	0.97***	0.83***
$A \cdot I_{30}^2$	0.98***	0.82***	0.98***	0.81***
$E \cdot I_{30}$	0.92***	0.77***	0.97***	0.82***
$E \cdot I_{30}^2$	0.98***	0.82***	0.98***	0.79***
$A \cdot E \cdot I_{30}$	A E I ***	***	0.94 ***	0.70 ***

[a] n = number of measurements; ** significant at $P<0.01$, *** significant at $P<0.001$.

3.2 *Erodibility of Inceptisols in the tropical Andes*

3.2.1 Long-term changes in erodibility

Continuously clean tilled fallow plots were established in 1986 at Quilichao and in 1987 at Mondomo. Table 6 presents yearly empirically determined K factors and soil losses.

In general, erodibility is classified as low to moderate in Quilichao and low in Mondomo according to Foster et al. (1981). Erodibility increased with number of years under cultivation. Increase in erodibility was faster in Mondomo; the K-factor already reached a value of 0.01 in the second year after establishment compared to a value of 0.004 in Quilichao. This may be due to the better aggregate stabilizing effect of the original grass vegetation in Quilichao as compared to that of the bush fallow in Mondomo.

In the third year erodibility increased sharply in Quilichao and continued to increase gradually until 1991-92. In Mondomo, the resistance of the soil against erosion diminished only slightly from the second year on.

In continuation of the present study, Castillo F. (1994) reported K-factors for the time period February 1992 to January 1993 (Tab. 6). Under very low rainfall regimes (1100 mm in Quilichao, 1200 mm in Mondomo) K-factors were as low as 0.005 in Quilichao and 0.01 in Mondomo. Corresponding soil losses were 35.2 t ha^{-1} (in 21 soil loss events) and 51.0 t ha^{-1} (in 19 events), respectively.

Table 6. Soil losses [a] and USLE erodibility factors (K [b]) at Quilichao and Mondomo, measured directly from permanent bare fallow plots.

	QUILICHAO			MONDOMO		
time period	*K-factor*	--- *soil loss* ---		*K-factor*	--- *soil loss* ---	
	plots established in 1986			plots established in 1987		
		t ha^{-1}				
1986-87	n.d.	n.d.	n.d.	-	-	-
1987-88 [c]	0.004		(49.2)	0.003		(44.5)
1988-89 [c]	0.013		(196.5)	0.010		(310.6)
1989-90 [d]	n.d.		(165.0)	n.d.		(190.0)
1990-91	**0.015**	**134.5**	**(143.7)**	**0.011** [e]	**89.3**	**(227.9)**
1991-92	**0.018**	**204.4**	**(222.6)**	**0.012**	**62.6**	**(160.1)**
1992-1993 [f]	0.005	35.2	(33.1)	0.010	51.0	(130.0)
plots established in 1991:	*K-factor*	*soil loss*				
1991-92	0.0014	15.0				

n.d. = not determined

[a] Corrected by the USLE S-factor (in brackets: original uncorrected values).

[b] K in t ha h ha^{-1} MJ^{-1} mm^{-1}; to obtain U.S. Customary units, K-factors (SI-units) are divided by 0.1317 (Foster et al., 1981).

[c] Data from Reining (1992).

[d] Soil loss data for 1989-90: Quilichao (CIAT, 1991a); in Mondomo losses were estimated (Cadavid, personal communication).

[e] For maintenance of the raingauge in 1990-91 in Mondomo, K-values could be calculated for 69 % of the soil losses only, excluding four heavy rainstorms.

[f] Measurements from February 1992 to January 1993 (Castillo F., 1994).

Three new permanent bare fallow plots were established in Quilichao in direct vicinity of the other bare plots in 1991. In the first year, after clearing the original grass sod (mainly *Paspalum notatum*), the K-value was as low as 0.0014, which is about 8 % of the erodibility of a plot six years under permanent bare fallow conditions and about 47 % of the erodibility of a bare plot in Mondomo in the first year after bush fallow clearing (Table 6). Roose (1980) reported increasing K-values of bare fallow plots three to five years after clearing a bush fallow, before they reached a near constant value. He suggested a decrease in organic matter content to be responsible for the increase in erodibility. Under permanent bare fallow conditions, organic matter content decreased from 7.3 % to 5.4 % in Quilichao in four years (CIAT, 1991b). Contents exceeding 4 % are believed not to decrease erodibility significantly (Young, 1976; cit. by Jansson, 1982). The stability of aggregates is generally greater under grass than under bush fallow (Aina, 1979). Probably after two years of bare fallowing the soil stabilizing effects of grass roots, their residues and microbiological activities during their descomposition decrease dramatically and lead to a strong increase in erodibility in Quilichao. It is unclear, however, if this suggestion is valid for well-aggregated soils of the tropical Andes. Adimihardja (1989) found no increase in K after clearing, for five of nine Indonesian soils studied, among them a Humitropept and a Dystropept, and stated that a decrease in organic matter content was not the only responsible factor for changes in erodibility.

3.2.2 Seasonal changes in erodibility and soil loss

Long-term measurements to give reliable K-values for a specific soil cannot be overemphasized (Röhmkens, 1985). USLE soil erodibility factors are long-term averages, including a wide range of storm sizes and antecedent moisture conditions. Erodibility is a dynamic property and varies among years, seasons and storms. Mutchler and Carter (1983) and others reported seasonal variation of the K-factor, in the USA: highest in winter to early spring and lowest in summer to early fall. Seasonal (dry, humid) variation of K-factors at our study sites are shown in Fig. 19 and results from three replications in Quilichao are given in Table 7.

Table 7. K-factors of dry and wet seasons in Quilichao during 1990 - 1992.

Season	Rep.1	Rep.2	Rep.3	average
Dry	0.012	0.011	0.013	0.012
Wet	0.015	0.023	0.020	0.019

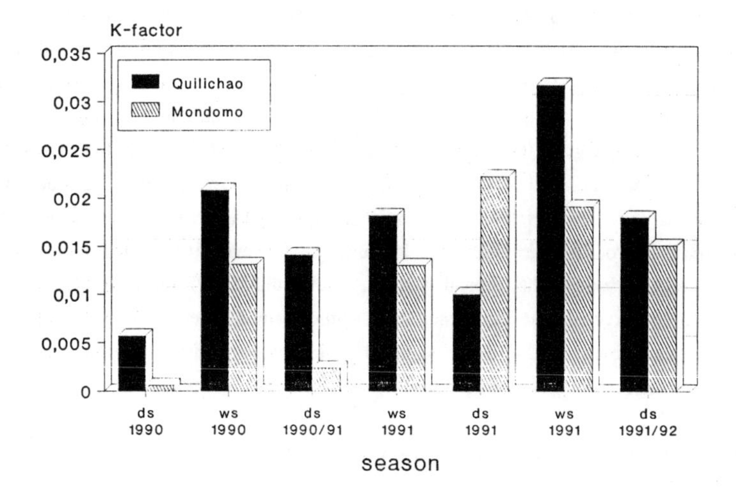

Figure 19. Seasonal Erodibility factors (K) in Quilichao and Mondomo.[a]

[a] dry seasons (ds): June 15 to September 31 and December 15 to February 28. wet seasons (ws): October 1 to December 14 and March 1 to June 14.

Erodibility is generally greater in the wet season, which may be explained by more frequent rain events of higher erosivity, higher moisture contents of soils and lower infiltration rates, and consequently higher water and soil losses. This finding is confirmed by Roose (1980) in West Africa.

Monthly K-factors are significantly correlated with amount of soil lost in Quilichao ($r = 0.86$) and in Mondomo ($r = 0.89$). Similarly, Ulsaker and Onstad (1984) found that K-factors depended on the magnitude of soil lost. Fig. 20 shows K-values for different soil loss ranges in the study period from April 1990 to April 1992. While there is a gradual increase of erodibility until 20 t ha^{-1}, followed by a sharp increase in heavy erosion events in Quilichao, in Mondomo no increase could be observed between 1 and 20 t ha^{-1}; above that amount erodibility stayed nearly the same. Only one event of soil loss of about

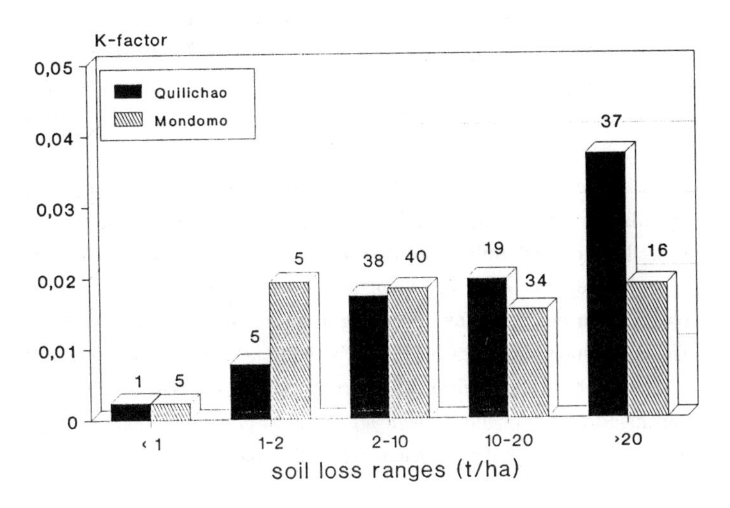

Figure 20: Amount of soil loss and corresponding erodibility factors (K) in Quilichao and Mondomo and the percentage contribution of soil loss in a certain range to total losses (April 1990-April 1992).

23 t ha^{-1} occurred here, but three heavy erosion events were observed in Quilichao; each with about 40 t ha^{-1}. Perhaps these large differences in soil losses during single events caused the very high K-factors at Quilichao and a smaller erosion susceptibility at Mondomo in events > 20 t ha^{-1}. Castillo F. (1994) reported higher soil losses and K-factors in Mondomo than in Quilichao in the study period February 1992 to January 1993 (Table 6). In Mondomo soil losses exceeded 10 t ha^{-1} in three, in Quilichao in two events only. An extremly high erodibility (K = 0.087) was found in one event in Mondomo, probably caused by a high antecedent moisture content of the soil.

One plot in Quilichao lost appreciably less soil in both years. In this plot some subsoil exposure could be observed, suggesting that with progressing topsoil loss, underlying horizons might show a lower erodibility than fertile topsoil. Additionally some gravel appeared (less than 3 % soil cover) and could have produced some erosion diminishing effect (Epstein et al., 1966). The erodibility of this plot was found to be lower in the wet season only (Table 7).

3.2.3 USLE calculation of K-factor

Properties of surface soils (0-10 cm), profile permeabilities and predicted K-factors, computed with the USLE-nomograph, are presented in Table 8. Predicted K-factors were less than measured values, in Quilichao more than in Mondomo. The nomograph thus appears not to allow the calculation of soil erodibility in the South Colombian Andes with sufficient accuracy.

Studies in the USA on the erodibility of clay subsoils (Röhmkens et al., 1975) and Minnesota topsoils (Young and Mutchler, 1977) revealed an appreciable underestimation of K-values for well aggregated soils. This is believed to be due to the mass removal of detached surface aggregates. After drying of the surface, soil aggregate formation was commonly observed on our bare plots, overlying a thin, 2 to 4 mm thick crust. The dry aggregate size distribution showed a negligible proportion belonging to sizes > 5 mm, compared to 15 to 20 % in the original soil under the soil crust. High erosion losses were always accompanied by rill formation, where aggregates or primary particles of bigger sizes can be transported over larger distances than by interrill erosion.

Table 8. Surface soil properties (0-10 cm) of permanent bare fallow plots, profile permeability and USLE-nomograph predicted K-factors in Quilichao and Mondomo.

	QUILICHAO	MONDOMO
	%	%
texture [a] (Pipette-method)		
clay (<0.002 mm)	75.03	64.14
silt (0.002-0.05mm)	15.68	21.15
v.f.sand (0.05-0.1)	2.13	1.51
f.sand (0.1-0.25mm)	5.55	6.16
rest sand (0.25-2.0mm)	1.61	7.04
organic matter (%) [b]	5.8	4.8
structure code [c]	3	3
profile-permeability class [d]	2	2
Nomograph K-factor	0.0033	0.0056

a soil samples for texture determination were taken in 1991 between the first and the second cropping period. Other parameters (OM, structure, permeability) are two year averages.

b 4 % org. matter was used for calculations (Young 1976; cit. by Jansson, 1982).

c Medium or coarse granular; determined by dry sieving (sieves with diameters of 10, 5, 2 and 1 mm).

d Final infiltration rates: Quilichao 6.9 cm h^{-1}; Mondomo 7.1 cm h^{-1}.

Röhmkens (1985) suggested that in aggregated soils, a structure term would better predict K-values than the texture term M of the original USLE-nomograph. El-Swaify (1977) suggested to include a measure for the contributions of amorphous or oxidic constituents to structural stabilities in the K-prediction. Trott and Singer (1983) in their study on Californian range and forest soils stressed the importance of clay mineralogy in soil erodibility. Certain sesquioxides (pyrophosphate- and oxalate-extractable Fe + Al) combined with a kaolinitic clay mineralogy appeared to act in reducing erodibility in otherwise erodible soils. Stability of aggregates showed highest negative correlation with soil loss in laboratory tests with Japanese Andosols (Egashira et al., 1983) and B-horizons of Ultisols (Egashira et al., 1986), both well aggregated soils with high clay contents.

3.3. *Conclusions*

The potential of soil erosion in the South Colombian Andes is high. Even on slopes of moderate gradients all fertile topsoil of permanent bare fallow plots may be lost within a decade. Soils of much steeper slopes are often cropped: planting time usually coincides with the onset of the rainy season and large rainstorms of high intensities may erode poorly covered soils to catastrophic extents within a few years. Precipitation in Mondomo was far below the long term average, indicating the possibility of much higher losses in 'normal' years.

For the two-year study period correlations between soil loss and different erosivity indices were generally higher in Mondomo. Here, indices with the highest accuracy of prediction were *EIA* (including runoff amount) and the products of the R-factor and rainfall amount (both $r^2 = 0.86$), I_{30} and rainfall amount ($r^2 = 0.81$), and the R-factor itself ($r^2 = 0.79$). In Quilichao, *EIA* by far correlated best with soil loss; it explained about 90 % of the soil loss variations. However, due to technical problems with the runoff infrastructure, this result should be analysed with caution. The USLE R-factor can be considered as a promising erosivity index in 1990 to 1992. These results are in agreement with the findings of Reining (1992) on the study sites in 1988-89. However, long-term investigation is necessary to obtain reliable information on the applicability of R as an erosivity index.

On lower slopes in Quilichao, interrill erosion probably contributes to a larger extent to soil loss. The products of either rainfall amount times the square of I_{30} or kinetic energy times the square of I_{30} (both $r = 0.86$) and the square of I_{30} alone ($r = 0.85$) yielded high

correlation coefficients. Compared to the $A \cdot I_{30}$ index, the inclusion of a simple antecedent moisture term, improved correlations from r = 0.79 to 0.84 in Quilichao only.
Erodibility measured on permanent clean tilled fallow plots under natural rainfall are classified as low to moderate in Quilichao and low in Mondomo. K-factors compare well with those of many Oxisols and Ultisols in the humid tropics. Highly significant positive correlations between K-values and the amount of eroded soil were found. The USLE nomograph is not applicable to estimate K-factors on Inceptisols, such as those found in the South Colombian Andes. Nomograph predicted values greatly underestimated measured K-values from continuous fallow plots, especially in Quilichao.

Part 4 *SOIL EROSION AND PRODUCTIVITY OF CASSAVA CROPPING SYSTEMS*

Cassava, the fourth most important staple food crop in the tropics in terms of joules produced for consumption (Cock, 1985), is grown mainly on marginal soils and often on steep slopes. It tolerates harsh climatic and edaphic conditions: water stress, low pH and nutrient status, high Al-saturation and Al-content, where other crops such as maize and beans suffer or even fail. Frequently, cassava takes the last position in the crop rotation; on very degraded soils it is the last option for the smallholder to grow an annual crop without the application of costly lime and fertilizers.

In southern Colombia, throughout the Departmento Cauca, cassava is grown for home consumption and local starch production on slopes from 10 to 60 %, in extreme cases up to 100 %.

Wide spacing and slow development of soil cover as a consequence of low soil fertility, ploughing for soil preparation, clean weeding by shovel or hoe together with cassava planting at the onset of the rainy season can lead to excessive soil losses and in a few decades to a complete loss of the soil resource.

Sound soil and crop management is the best method of erosion control (Lal, 1980). Well fertilized plants achieve a faster soil cover (CIAT 1984); simultaneous intercropping of cassava with a fast growing crop of short duration such as maize or beans (Aina et al. 1976), or with *Stylosanthes hamata* (El-Swaify et al, 1988) diminish soil and water losses. Application of mulch protects soil from direct raindrop impact, improves water infiltration and hence diminishes erosion (Lal, 1976a; Lehle, 1986). Minimum tillage (Lal, 1976b, Ryan, 1986), contour ridges and contour grassbarriers can reduce erosion losses effectively (Fournier, 1967; Bharad and Bathkal, 1990; Margolis et al., 1991; Reining, 1992).

On the other hand, a number of authors have reported increased soil losses when cassava was intercropped with annual grain legumes such as cowpea, beans and groundnuts (Howeler, 1985a; Reining, 1992; Cadavid, personal communication). This is generally attributed to the additional intensive soil preparation to provide a favorable seedbed for the grain legumes.

On more eroded soils cassava produces reasonable yields, when soil preparation is done by plowing, which may result in high soil losses. On newly cleared old bush fallow or less degraded better structured soils, minimum tillage - only planting holes were prepared - were a very effective means to control soil loss and to produce high yields (Cadavid and

Howeler, 1987). On acid soils of low fertility, generally, intercropped cassava produced less than when grown alone (CIAT, 1978; Wilaipon et al., 1981; Hegewald, 1990); grain legumes not always compensated for the yield losses caused by competition for water, nutrients and light (Reining, 1992). In contrast, Nitis and Sumatra (1976) reported results from Bali which show a 17 % yield increase of cassava when intercropped with *Stylosanthes guianensis* as compared to sole cropping.

4.1 *Soil loss and runoff*

4.1.1 Distribution and erosivity of rainfall in the cropping seasons 1990-91 and 1991-92

Monthly distribution of precipitation and its erosivity is shown in Fig. 18a and 18b. The R-factor of the Universal Soil Loss Equation (Wischmeier and Smith, 1978) was significantly correlated with soil losses from standard bare fallow plots ($r = 0.73$ to 0.93; $P<0.001$).

Less rain fell in Quilichao in the first cropping season with 1450 mm than in the second one with 1654 mm. R-factors per cropping season of 341 and 373 days were 8612 and 11078 MJ mm ha^{-1} h^{-1}, respectively. Rainfall in the first three months after cassava planting was 16 % less (236 mm) than in the second cropping season, but the erosive power of rainfall with a R-factor of 1881 was about twice as high. Four heavy rainstorms with high intensities in November and December 1991, February and April 1992 caused the greater overall erosivity in the second season: They accounted for 38 % of the year's R-factor.

Rainfall in Mondomo of 1126 mm in 1990-91 and of 1275 mm in 1991-92 was far below average. R-factors per cropping season of 341 and 379 days were 6135 and 5142 MJ mm ha^{-1} h^{-1}, respectively. No significant erosive rain events occurred in the first cropping season after cassava planting until October 1990. Erosivity in the second period was low until November 1991. In the dry season in February 1992, heavy rainfall of high intensity occurred; one single rainstorm contributed 24 % to the year's total erosivity.

4.1.2 Soil cover development

P.phaseoloides was seeded 80 days before planting cassava in the first cropping season in Quilichao. Total soil cover (cassava, legume, mulch and weeds) of this cropping system

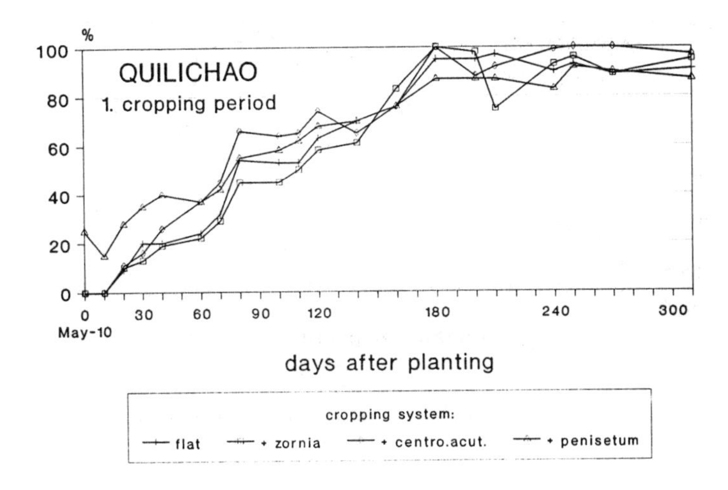

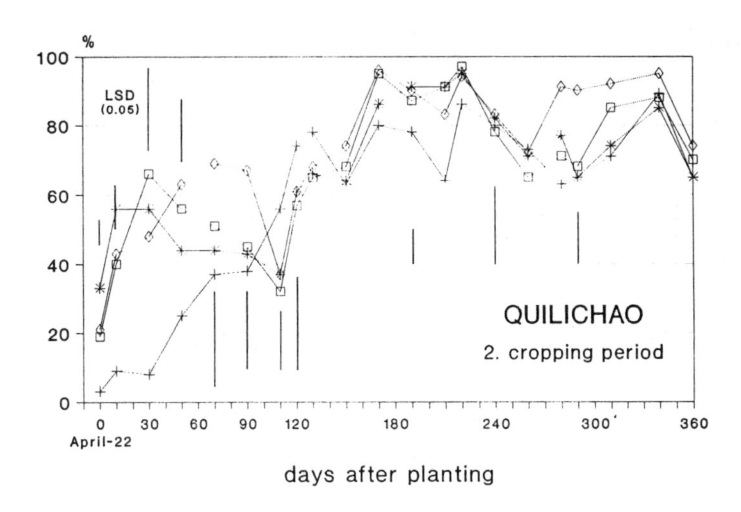

Figure 21a: Development of total soil cover (cassava+mulch+weeds, %) in monocropped and intercropped cassava in Quilichao in the first and second cropping period.

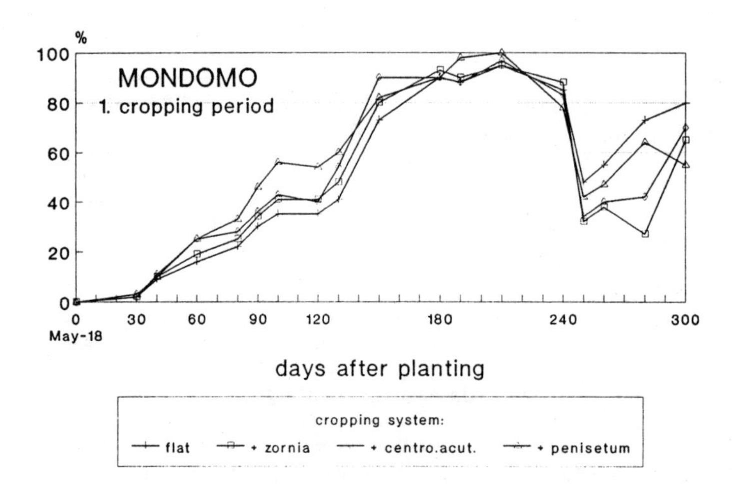

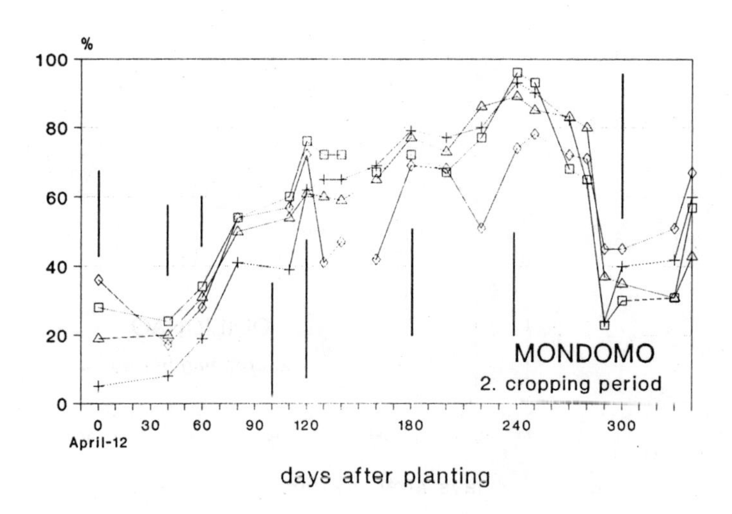

Figure 21b: Development of total soil cover (cassava+mulch+weeds, %) in monocropped and intercropped cassava in Mondomo in the first and second cropping period.

was significantly greater ($P<0.05$) than sole cassava in flat cultivation, contour ridging or the other intercropping systems until 50 days after planting (Fig. 21a). The soil cover of cassava intercropped with *C.acutifolium* developed fastest and from 90 days on, no significant differences among cropping systems were found. In the second season total soil cover of all cassava-forage legume intercropping systems was greater than traditional flat cassava sole cropping in the first three months after planting, the most critical period for erosion in cassava cropping: Averaged over six measuring dates cassava with *P.phaseoloides*, *Z.glabra* and *C.acutifolium* reached a soil cover of 46, 46 and 52 %, respectively, whereas cassava grown on the flat covered only 20 %. From the end of September until the middle of December cover of intercropped systems was again greater, in November reaching significant differences ($P<0.05$) in cassava with *C. acutifolium* and *P.phaseoloides* and again from middle of January until end of February in cassava associated with *C.acutifolium*.
Soil cover developed more slowly in Mondomo (Fig. 21b). In the first cropping period flat cultivation and intercropping did not show any significant differences. In 1991-92 all intercropping systems reached a greater degree of soil cover in the first four months than traditional cropping. After cutting both Centrosema varieties in November, total coverage fell significantly below sole cropped cassava, but recovered fast and after harvest at the end of January, soil cover of this intercropping systems was greatest.
Development of the legumes is shown in Table 9 for Quilichao and Mondomo for the first 120 days after seeding in 1990-91. Establishment of forage legumes was quite difficult. In Quilichao, one heavy rain the day after seeding washed soil and many seeds away, especially the small seeded *Z.glabra*.

Table 9. Development of soil cover in % of total plot surface of forage legumes intercropped with cassava in Quilichao and Mondomo from May to September 1990.

Days after cassava planting:		**0**	**30**	**60**	**90**	**120**
				% cover		
QUILICHAO	*P.phaseoloides*	33	29	36	45	44
	Z.glabra	0	1	4	14	19
	C.acutifolium	0	3	15	37	45
MONDOMO	*Z.glabra*	0	0	0	2	2
	C.acutifolium	0	0	3	8	10
	C.macrocarpum	0	0	3	11	15

Dry weather conditions made irrigation necessary. Legumes had to be reseeded. Higher temperatures and a better water supply favoured legume establishment in Quilichao.
C.acutifolium and *P.phaseoloides* climbed into the cassava and had to be cut; regrowth in the beginning of the rainy season in October was very fast and cutting had to be repeated two months later. *Z.glabra* was cut, nearly seven months after planting, the first time at 70 % soil cover. Once established, it formed a perfect sod, effectively diminishing erosion. It reseeded freely and had to be weeded around the cassava plants. In one of the plots with this association, soil cover by cassava reached 95 % and suppressed the regrowth of *Z.glabra*; legume cover stayed low until the end of the next cropping season.
In Mondomo legume development in the first four month of 1990 was very slow (Table 9). Average temperature was at about 19°C, and rainfall in the establishment period was scanty (Fig. 18a and 18b). Legumes in cassava plots had to be reseeded and irrigated three times. Then abundant rains fell in October and both Centrosema varieties grew vigorously. They reached a soil cover of 60 % at the end of October and climbed into cassava which made cutting necessary. *Z.glabra* could not be established satisfactorily in the first cropping season.
A heavy white grub attack in December and January reduced soil cover of legumes and cassava sharply, many plants died or shed their leaves. Damage was heavier in *C.acutifolium* than in *C.macrocarpum*. In February 1991, reseeding of *Z.glabra* was successful. Rainfall conditions were favorable and in August cover reached 35 %. *C.macrocarpum* seems to be better adapted to cooler temperatures than *C.acutifolium*. It had to be cut four times in the second season for its more aggressive growth habit. Regrowth was always faster and production higher than *C.acutifolium*. White grub attack occurred again in January 1992 with some damage to *C.acutifolium*, but heaviest to *Z.glabra*, which was nearly completely lost after cutting. *C.macrocarpum* did not show any significant damage.
Forage legumes were cut regularly and material was carried out of the plots. Farmers are not likely to adopt erosion control practices, if no direct benefit is obvious. This is because soil conservation generally involve higher costs, increase labour demand, reduce cropping area (terraces, contourstrips), require technical skill and often cause lower yields of field crops in the short run. In the case of this study, forage legumes were expected to be used as animal fodder on smallholder farms. Leaving cut forage legumes as a mulch in the cassava fields probably would have produced different results, favouring more the intercropping systems.

4.1.3 Soil losses as influenced by crop management

Fig. 22 shows accumulated soil losses in Quilichao and Mondomo in the first and second cropping period. Values were corrected for corresponding Slope x Length-factors (LS) of the USLE (Wischmeier and Smith, 1978). They correspond to losses from slopes with a gradient of 9 % and a length of 22.1 m. The validity of the USLE L- and S-factors are assumed, but may not be true for the climatic and edaphic conditions studied here. However, corrections do not alter the qualitative differences in soil losses among treatments.

On this basis soil losses in Quilichao were significantly ($P<0.05$) greater than in Mondomo. Reasons are the greater erosivity (32 % greater in 1990-91 and 132 % greater in 1991-92) and a greater erodibility of soils (Reining, 1992), probably together with the soil preparation done by rotovator, leading to a less rough surface of cropped plots than with preparation by oxen plough in Mondomo.

Maximum soil losses in the study region tolerated to permit a sustainable high level of crop productivity (T-values) were estimated by Reining (1992) and this author to range between one and two tons per ha and year. Seeding and reseeding operations of legumes caused soil disturbance and aggregate disintegration and led to very high soil losses even on flat slopes. *C.acutifolium* had to be reseeded twice, *Z.glabra* three times. *P.phaseoloides* was already established at planting time (33 % soil cover; not shown in Fig. 21), but not uniformly distributed, making reseeding necessary. The seedbed for cassava flat planting showed a higher degree of roughness than in intercropped plots. Soil losses increased with increasing degree of 'human' traffic on the plots for seeding and reseeding operations in the order: cassava associated with *P.phaseoloides* (14.2 t ha^{-1}), *C.acutifolium* (17.0 t ha^{-1}) and *Z.glabra* (27.2 t ha^{-1}). In traditional flat cassava planting, 7.1 t ha^{-1} was lost. Soil loss from the cassava-*Z.glabra* treatment was significantly ($P<0.05$) greater than with flat planting, contour ridging or grass barriers.

The erosion risk was greatest in the first three to four months of the cassava cropping cycle. In the first month only one heavy and very intense rainfall occurred and caused about 80 % of the total annual soil loss in flat cropped cassava and when associated with *Z.glabra* and *C.acutifolium*. About 70 % was lost in this time period when cassava was intercropped with *P.phaseoloides*. Plots under permanent bare fallow, which generally represents the most erodible condition, lost 13.8 t ha^{-1} in this time period.

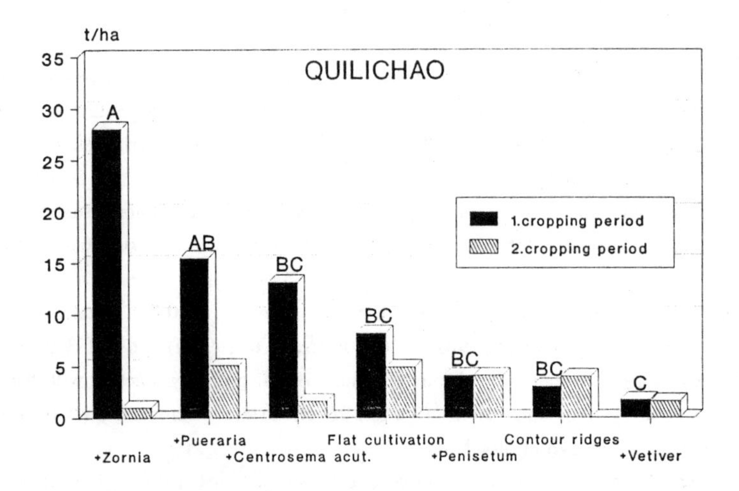

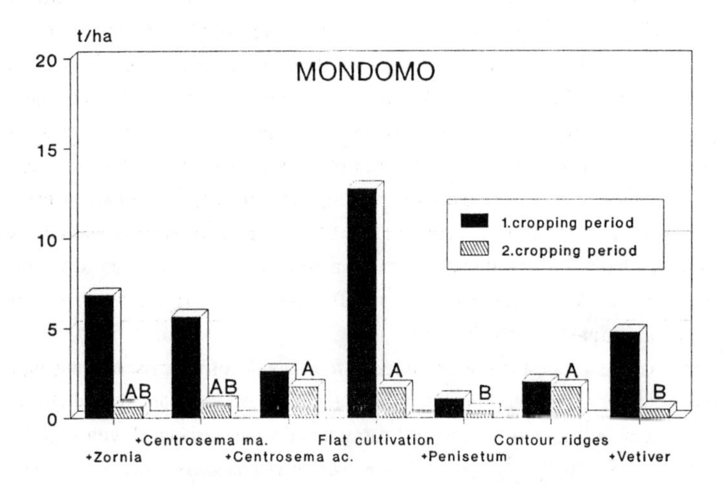

Figure 22: Accumulated dry soil losses in different cassava cropping systems in Quilichao (above) and Mondomo (below) in the first and second cropping period.

However, soil losses in the cassva-*Z.glabra* plots exceeded those of bare fallow by 60 % indicating the high susceptibility of soils at Quilichao to disturbing activities such as "human traffic", which was inevitable on intercropped cassava plots. Intercropping with *C.acutifolium* and *P.phaseoloides* required less "human traffic" for reseeding and thus soil losses were 5 and 30 %, respectively, below those of the bare fallow plots. Solecropped, flat cultivated cassava lost 60 % less soil than the control.
Best erosion control was achieved with contour ridging and the use of grass barriers planted to elephant grass and vetiver grass which reduced soil losses to 3.7, 4.7 and 1.7 t ha^{-1}, respectively. The effectiveness of vetiver grass barriers was outstanding altough it has to be considered that this system was established one month after the other cropping sytems escaping the early erosive events. Furthermore, these plots were newly established on an area with previous long term stable *Paspalum notatum* grass cover. They showed excellent physical characteristics (aggregate stability, infiltration) and a high amount of grass root residues. The other plots had been in continuous cassava cropping for four years in Quilichao and three years in Mondomo.
A rainstorm of 31 mm in July 1990 accompanied by heavy winds caused appreciable losses in the intercropping systems only; greatest soil losses were observed with *Z.glabra* and somewhat smaller losses with *C.acutifolium* and *P.phaseoloides*. This tendency could be seen until the onset of the rainy season in October 1990. Until October no significant negative correlation was found between soil cover and soil loss, stressing the predominance of erosion enhancing cultivation and establishment practices over erosion reducing cover effects in the first months of the cropping cycle. Afterwards, soil cover by cassava, weeds and legumes increased rapidly as a response to good water supply and significant negative correlations with soil loss were found in November ($r = -0.59$, $P<0.01$). Cassava on contour ridges was an effective means to control erosion on these gentle slopes and well aggregated soils.
Previous research done by CIAT confirms greater soil losses in cassava intercropping systems than in sole cropping in the study region. Howeler (1985a) reported soil losses in the Mondomo area on a 30 % slope of over 40 t ha^{-1} with solecropped cassava in 14 months, whereas in cassava intercropped with cowpea more than 50 t ha^{-1} were lost, representing a topsoil loss of over 0.5 cm. On sites used for the present study Reining (1992) and Cadavid (personal communication) found that intercropping with grain legumes caused greater soil losses than traditional solecropping and attributed this to the additional intensive soil preparation to provide a favorable seedbed for the grain legumes.

In the second cropping period in Quilichao soil losses were not significantly different among the cropping systems. Intercropping of cassava with *Z.glabra* resulted in very small soil losses of 1 t ha^{-1}, with *C.acutifolium* 2.2 t ha^{-1} were lost. With the exception of one repetition of the cassava-*Z.glabra* association, where a high cassava soil cover suppressed the regrowth of the forage legume after cutting, both legumes formed a nearly perfect sod covering the entire plot. When cut, a dense stand and a mulch of fallen leaves provided an excellent protection against raindrop impact and runoff. Soil losses recorded for cassava with *P.phaseoloides*, solecropped on the flat, on contour ridges and between barriers of elephant grass were 4.1, 4.0, 4.6 and 4.6 t ha^{-1}, respectively. Reasons for these losses may be the non-uniform plant stands and distribution within the plots, and, in the case of legumes other than *C.acutifolium*, the development of less trailing stems. Gaps in the stand of elephant grass contour strips and poor cassava growth on the first two rows next to the grass (Fig. 23) as a consequence of competition for water, light and nutrients led to concentrated runoff and somewhat greater losses than traditional flat cropping.

The erosivity of the rainfall in the first three months after planting of the second cassava crop was 50 % lower than in the preceding year (Fig. 18a and 18b) and caused very small losses. These rainfall conditions were not really a test for the effectiveness of erosion control of intercropped systems in the second year: Soil preparation by minimum tillage and legume cover would have prevented any excessive erosive effect of much higher precipitation. When highly erosive rains occurred in November and December 1991, February and April 1992, soil cover of plants, mulch and weeds reached 80 % (averaged over all treatments), which was large enough to prevent excessive losses in all cropping systems. Nevertheless, these losses were appreciable; only uniformly distributed and dense stands of legumes such as *Z.glabra* and *C.acutifolium*, or dense gapless contour grass barriers such as those with *V.zizanioides* were able to keep soil losses within tolerable limits (on the basis of 9 % slope and 22.1 m length).

Results of soil losses from ridge cropped cassava are misleading. Losses were relatively high (4.6 t ha^{-1}): the ridges were only sparsely vegetated with weeds and soil cover of the cassava variety CM 523-07 was less than the year before with CM 507-37. Furthermore, the steep slope of the ridge next to the collection channel was nearly completely exposed to the rains. Thus, soil was lost from the first ridge and cannot be considered as soil lost from the entire plot. Although there was some flattering of the ridges over the month, no breakthrough of ridges occurred at any time during the growth cycle.

In Mondomo no significant differences between soil losses among cropping systems were found in the first cropping period (Fig. 22). No losses occurred until September; crops suffered from drought and at the initiation of the rainy season in October, soil cover was still low with 48 %, averaged over all treatments. Again soil loss was not significantly correlated with soil cover in October, but in November a significant negative correlation was found ($r = -0.58$; $P<0.05$). Flat cultivation of cassava led to greatest soil losses of 4.7 t ha^{-1} with 79 % of total losses recorded in October. This was followed by Cassava with *Z.glabra* registering soil loss of 2.7 t ha^{-1} (82 %), by cassava with *C.macrocarpum* producing soil losses of 2.1 t ha^{-1} (74 %) and by cassava with *C.acutifolium* where a soil loss of 1.3 t ha^{-1} (55 %) was produced. Soil losses from flat cultivated cassava would have been greater, taking into account, that sediments deposited at the lower end of one plot and did not reach the collection channel. Establishment operations for legumes did not cause greater losses than traditional cropping. Soil was probably more resistant to human traffic than in Quilichao and erosive rain events did not occur in the first four months after planting. High erosivity in November was more than offset by the rapidly developing soil cover. Over 80 % cover (averaged over all cropping systems) controlled erosion effectively. Cassava and contour lines of vetiver grass established on plots in bush fallow for the previous two years lost 3 t ha^{-1}. Elephant grass established well with a denser stand than in Quilichao; soil losses were lowest with 0.6 t ha^{-1}, followed by cassava on contour ridges with 0.9 t ha^{-1}. In the second cropping season soil losses were generally low (Fig. 22). Only the difference between cassava associated with *C.acutifolium* and with barrier strips of *P.purpureum* barriers reached significance with 1.0 and 0.2 t ha^{-1} soil loss, respectively.

These results are in accordance with those of Reining (1992) reported from the same study locations. He found best erosion control, when cassava was cropped on contour ridges. Contour grass barriers were only effective in Mondomo. In Quilichao, however, soil losses were similar to flat traditionally cropped cassava. Another very effective practice was the cropping of cassava with minimum tillage.

These data show that individual crops or cropping systems may differ substantially with regard to their influence on soil loss, depending on time and degree of soil cover provided. Soil cover and the condition of the soil itself is influenced by management such as tillage, crop selection, planting density, fertilization and other cropping practices. The relationship between soil loss and crop management is taken into account by the cropping factor C of the USLE, defined as the ratio of soil loss from land cropped under specific conditions to the corresponding loss from clean tilled, continuous fallow (Wischmeier and Smith, 1978). For the cropping systems under study, C-factors are given in Table 10. The results stress the importance of erosion control measures in early stages of the cassava development.

Table 10. Annual USLE C-factors of selected cropping systems in Quilichao and Mondomo, crop stage C-factors and corresponding R (erosivity, %) distribution in Quilichao.

Cropping systems:	QUILICHAO					MONDOMO
	crop stage periods [a]					
1990-91 cropping period	**1**	**2**	**3**	**4**	**annual**	**annual**
	C-factor					
flat cultivation	0.32	0.00	0.01	0.01	**0.05**	**0.07**
+*P.phaseoloides*	0.59	0.00	0.04	0.10	**0.13**	-
+*Z.glabra*	1.44	0.00	0.03	0.01	**0.21**	**0.04**
+*C.acutifolium*	0.79	0.00	0.03	0.01	**0.13**	**0.02**
+*C.macrcarpum*	-	-	-	-	-	**0.03**
% R-distribution	21	11	50	18		
1991-92 cropping period						
	C-factor					
flat cultivation	0.07	0.02	0.02	0.03	**0.02**	**0.01**
+*P.phaseoloides*	0.07	0.02	0.01	0.08	**0.02**	-
+*Z.glabra*	0.04	0.01	< 0.01	0.02	< **0.01**	< **0.01**
+*C.acutifolium*	0.07	0.02	0.01	0.03	**0.01**	**0.02**
+*C.macrocarpum*	-	-	-	-	-	< **0.01**
% R-distribution	8	12	64	16		

[a] Crop stage periods as defined in Part 2 "Materials and Methods".

4.1.4 Runoff losses in cassava cropping systems

Runoff rates (% rainfall lost as runoff) correspond to slopes with 7 - 13 % gradient in Quilichao and 12 - 20 % in Mondomo. Runoff rates were low, however, some runoff was lost in heavy rainstorms (Table 11). In the first cropping season measurements in Quilichao began in July 1990 after high and erosive precipitation in May, when the major soil losses occurred. Somewhat higher water losses occurred in cassava intercropped with *Z.glabra*: 6.3 % of rain was lost in the first, 7.2 % in the second season. This increase in runoff may have been caused by soil disturbing activities affecting infiltration following human traffic during cassava/legume establishment and during minimum tillage cassava planting in the existing legume sward in 1991-92.

Higher runoff rates were found in the first season on the steeper slopes in Mondomo than in Quilichao, again highest in the cassava-*Z.glabra* association (8.2 %) and lowest, when cassava was grown on contour ridges (3.9 %). During the second cropping period plots at Mondomo received less rainfall of lower erosivity than Quilichao leading to markedly lower rates of runoff in the former compared to the latter location. Lowest rates were observed with life barriers (*P.purpureum* 2.3 %, *V.zizanioides* 2.2 %) and *Z.glabra* (2.4 %). In none of these situations, differences in runoff rates among cropping systems were significant over both years.

Table 11. Runoff rates (% rain lost as runoff) of different cassava cropping systems in Santander de Quilichao (7 - 13 % slope) and Mondomo (12 - 20 % slope) in the cropping periods 1990-91 and 1991-92.

	QUILICHAO		MONDOMO	
Cropping system	1990-91 [a]	1991-92	1990-91 [b]	1991-92
	----------runoff rates % ----------			
contour ridges	3.4	3.8	3.9	3.4
flat cropping	3.6	5.0	6.4	3.9
+*P.phaseoloides*	3.4	5.7	-	-
+*Z.glabra*	6.3	7.2	8.2	2.4
+*C.acutifolium*	3.0	5.5	6.6	2.9
+*C.macrocarpum*	-	-	6.2	3.1
+*V.zizianioides*	3.5	4.9	5.6	2.2
+*P.purpureum*	3.9	5.6	5.2	2.3

[a] 9 months measurements
[b] 7 months measurements

4.2 *Forage and cassava productivity*

4.2.1 Forage legumes and contour grasses

Over both years in Quilichao, highest production of dry matter was achieved by *C.acutifolium*, producing similar yields in both cropping seasons (Table 12). One reason may be a higher shade tolerance of this legume (CIAT, 1988). Production of *P.phaseoloides* and *Z.glabra* decreased by 35 % and 29 %, respectively, in the second season.

Elephant grass produced 5.1 t DM ha^{-1} of fodder in Quilichao and 7.3 t ha^{-1} in Mondomo averaged over both cropping periods on 25 % of the plot area. Vetiver grass, on 12.5 % of the plot area, established slowly, but doubled its production in the second year. Two-year average DM yields were 1.8 t ha^{-1} in Quilichao and 2.1 t ha^{-1} in Mondomo.

Table 12. Dry matter yields of forage legumes and contour barrier grasses (number of cuttings in parenthesis) in Quilichao and Mondomo in the first and second cropping period.

	QUILICHAO		MONDOMO	
	cropping period		cropping period	
legume/grass	1990-91	1991-92	1990-91	1991-92
	DM t ha^{-1}			
P.phaseoloides	2.9 (3)	1.8 (3)	-	-
Z.glabra	3.4 (2)	2.4 (3)	0.1 (1)	1.4 (2)
C.acutifolium	3.3 (4)	3.4 (3)	1.4 (2)	2.4 (3)
C.macrocarpum	-	-	1.8 (2)	3.7 4)
V.zizanioides	1.1 (2)	2.4 (3)	1.2 (1)	2.9 (4)
P.purpureum	5.6 (5)	4.6 (5)	8.0 (3)	6.5 (5)

4.2.2 Cassava root yield

Similar to results reported by Reining (1992), in the present study cassava on contour ridges and on the flat produced greatest root yields in Quilichao in 1990-91. In the second cropping period these cropping systems as well as cassava planted with vetiver grass barriers yielded best (Table 13). In the first period, differences among yields of the cropping systems were not significant, however, some tendencies could be noted. Yields of intercropping treatments were highly variable. Forage legumes seemed to depress cassava yields slightly. Yield reductions were 24 and 11 % when cassava was undersown with *Z.glabra* and *C.acutifolium*, respectively. These results compare well with those reported by CIAT (1978) and Hegewald (1990) in Quilichao for most cassava forage legume associations tested. *P.phaseoloides* already covered soil to 33 % when cassava was planted. Soil cover and plant height of cassava was lower than in all other treatments over the whole growing period and yield was 42 % less than traditional flat planting. Ten month old cassava with barriers of *V.zizanioides* had been planted on land previously under grass fallow and yielded well in spite of planting in the dry season.

Table 13. Fresh root yield of cassava in cassava cropping systems in Santander de Quilichao and Mondomo during the first 4-5 years of the experiments. [a]

	QUILICHAO[b]				MONDOMO[b]		
	cropping periods				*cropping periods*		
cropping systems:	1987 [c] -1989	1989 [d] -1990	1990-1991	1991-1992	1988 [e] -1989	1990-1991	1991-1992
	t ha^{-1}						
solecropping:							
contour ridges	30.7a	28.4	35.6a	23.3a	15.3a	15.4abc	13.4a
ridges downslope	28.3a	-	-	-	15.4a	-	-
flat cropping	31.9a	28.5	35.7a	22.7ab	19.7a	18.4a	13.5a
minimum tillage	7.7c	-	-	-	15.7a	-	-
mulching:	-	30.9	-	-	-	-	-
intercropping:							
+*P.phaseoloides*	-	-	20.8a [f]	16.0ab	-	-	-
+*Z.glabra*	-	-	27.1a	12.9b	-	17.5ab	10.6a
+*C.acutifolium*	-	-	31.8a	13.2b	-	18.2a	7.9a
+*C.macrocarpum*	-	-	-	-	-	11.4c	7.9a
+grain legumes	20.4b	22.1	-	-	16.8a	-	-
grass barriers:							
+*V.zizanioides*	-	-	28.6a [g]	23.5a	-	12.4bc	12.2a
+*P.purpureum*	30.2a [h]	24.4	23.6a	16.2ab	18.2a	12.8abc	11.0a

[a] Values with the same letters in a column are not significantly different from each other.
[b] In the cropping periods 1990-91 and 1991-92 cassava was harvested at an age of 11 months in Quilichao and in Mondomo at 8 and 9 months, respectively.
[c] Mean of two cropping periods; Cassava variety CM 523-07. Data from Reining (1992).
[d] Data from Cadavid (personal communication), CIAT, Santander de Quilichao, Colombia; cassava variety CM 507-37.
[e] Cassava variety M-Col 1522 ('Algodona'). Data from Reining (1992).
[f] *P.phaseoloides* was seeded 80 days before cassava planting.
[g] 10 months old Cassava.
[h] In the first two years only: *Paspalum notatum* as contour grass barriers.

In the second cropping period significant differences between cassava yields among cropping systems occurred. Greatest yields were achieved on contour ridges and with vetiver grass barriers. They were significantly greater than in the association with

C.acutifolium and *Z.glabra* ($P<0.05$). These legumes depressed cassava yields more than 40 %. Cassava was planted in minimum tillage (only planting holes were prepared) in the legume sward. The legumes competed from the beginning of the cropping cycle with cassava. Furthermore, minimum tillage has shown to be a soil preparation method, which decreases cassava yields on the study sites. Reining (1992) reported a root yield reduction in Quilichao of over 75 % which was attributed to the competition effect of the native grass sod. Reduction by *Z.glabra* would have been worse taking into account only two repetitions, where the stand of *Z.glabra* was excellent and cassava yields were decreased by over 57 %. Associated with *P.phaseoloides* yields were 30 % lower compared to sole cropped, flat planted cassava. Height of cassava (Table 14), when planted with legumes, was lower over the whole cropping period; height reductions in the intercropping systems coincided with their root yields.

In Quilichao, over both cropping periods highest harvest indices were found in cassava on contour ridges (0.80 and 0.75) and lowest, when cassava was intercropped with *P.phaseoloides* (0.64 and 0.66).

In Mondomo, cassava had to be harvested after 8 months in the first and after 9 months in the second period to avoid root losses by heavy attacks of white grubs, which explains the low yields (Table 13). Cassava on the flat and intercropped with *C.acutifolium* gave highest yields in Mondomo in the first cropping season. Reining (1992) reported highest yields when cassava was flat cropped. When intercropped with beans, yields were 15 % lower, however, no significant differences occurred among cropping systems. *Z.glabra* reached an average of 3.3 % soil cover, so this treatment was almost equal to flat planted, solecropped cassava. In the first 90 days after planting, only about 100 mm of rain was recorded so that cassava suffered from drought, especially, when cultivated on contour ridges. A third part of its stakes had to be replanted in June and may have caused decrease in yields. Roots of cassava, intercropped with *C.macrocarpum* were attacked by white grubs more than in any other cropping system. Pest damage as well as the very good development of *C.macrocarpum* may be the reason for the drastic reduction of cassava yields in the first year, when yields were 38 % below those of cassava grown on the flat. During the whole cropping period soil cover by cassava was least. Cassava cropped with contour barriers of elephant grass, produced 7 % less roots than flat cropping, when calculated on the basis of an area of 100 %.

In the second cropping period cassava on the flat and on contour ridges produced greatest root yields. The latter showed the highest harvest index in both cropping periods. Cassava with contour strips of *P.purpureum* produced 81 % of flat cropped cassava on 75 % of the area, and with contour strips of *V.zizanioides* the production was 90 % on 87.5 % area (Table 13). This positive yield effect is probably due to the higher planting density of cassava in the vetiver grass (11,111 plants ha^{-1}) and elephant grass (12,500 plants ha^{-1}) treatments as opposed to solecropped cassava (10,000 plants ha^{-1}). In the study area around Mondomo farmers generally plant cassava at higher densities (15 - 20,000 stakes ha^{-1}), than those used in the present study. This practice may be a good means to compensate to some extent for the yield losses by reduction in cropping area. Over the whole year soil cover and plant height (Table 14) of cassava associated with legumes was lower than when sole crop-flat planted, reflecting yield reductions of cassava, amounting to 21 % with *Z.glabra*, 41 % with *C.acutifolium* and *C.macrocarpum*. Some yield decrease may be attributed to planting with minimum tillage. Reining (1992) found yield reductions of about 20 % in Mondomo, when cassava was grown with minimum tillage compared to sole crop-flat planting, depending on the variety used.

Table 14. Development of height of cassava plants in Quilichao and Mondomo. Flat sole- and intercropped cassava in the second cropping period 1991-92.

		Days after planting				
Site	Cropping system	60	120	180	240	300
		cm				
QUILICHAO	flat solecropping	47 a [a]	98 a	123 a	139 a	133 a
	+*P.phaseoloides*	31 b	64 b	102 ab	122 a	122 a
	+*Z.glabra*	28 b	54 b	93 b	113 a	111 a
	+*C.acutifolium*	24 b	56 b	93 b	114 a	115 a
MONDOMO	flat solecropping	21 a	53 a	78 a	95 a	-
	+*Z.glabra*	20 a	48 a	66 a	89 a	-
	+*C.acutifolium*	19 a	44 a	62 a	82 a	-
	+*C.macrocarpum*	16 a	46 a	68 a	85 a	-

[a] Values with the same letters in a column are not significantly different from each other.

4.2.3 Competition effects on cassava

Elephant grass barriers competed strongly with cassava. The yield of the two rows next to the grass barrier was suppressed (Fig. 23). Calculating yields on an area of 100 % cassava

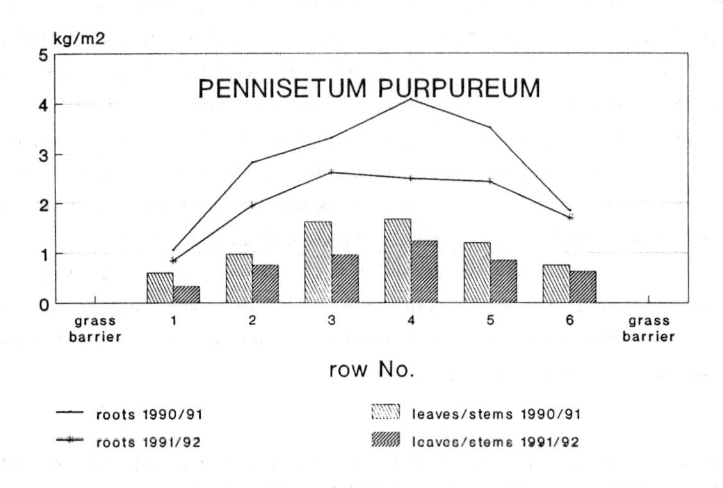

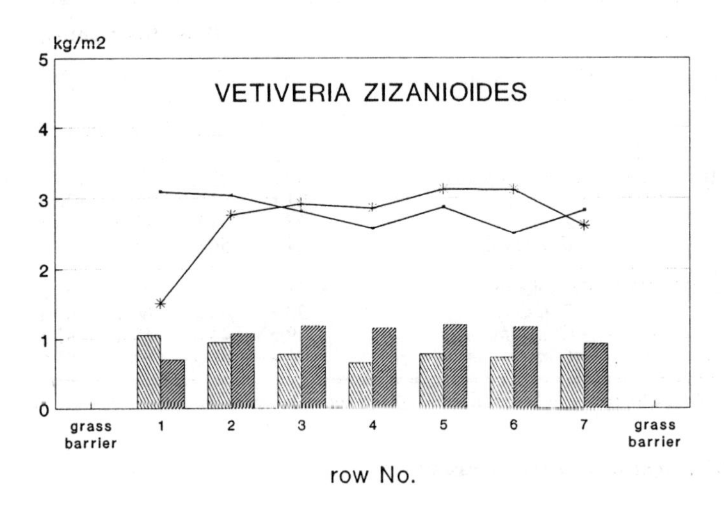

Figure 23: Yields of fresh roots and aboveground biomass of cassava (kg m^{-2}) between contour barriers of *P.purpureum* (above) and *V.zizanioides* (below) in the first and second cropping period.

shows that competition by elephant grass effectively decreased yields by 12 % in the first and 5 % in the second cropping period compared to flat planting without grass barriers. In the year of vetiver grass establishment, no yield decrease of the rows next to the grass could be noted. One year later competition occurred, but it seemed to be less pronounced than that imposed by elephant grass. One reason for this difference in competitivity of the grasses may be that the root system of vetiver grass extends more in vertical than in horizontal direction (Tscherning et al., 1995). Moisture and nutrient status is thought to be improved near the grass row with time, and requirements of vetiver grass for nutrients are low (Sethi, 1982).
Concentrations of nitrate N and exchangeable K of the soil between the grass barriers and the first adjacent cassava row (+competition) were significantly lower than in cassava without grasses (-competition; Table 15). Together with water stress, the competition for these nutrients may be responsible for the yield decrease. Critical K levels for a nearly optimum yield have been shown to be 0.17 - 0.18 cmol kg^{-1} soil (Howeler, 1985b).

Table 15. Concentration of nutrients in topsoils (0-20 cm) of plots with well established contour grass barriers 1991-92 adjacent (+competition) and distant (-competition) to the grass barrier.

Position	NH_4^+	NO_3^-	Bray-II P	K^+	Ca^{2+}	Mg^{2+}	Zn
	------	mg kg^{-1}	-----	----	cmol kg^{-1}	----	mg kg^{-1}
- competition	30.6	15.2	14.2	0.19	1.35	0.56	1.2
+ competition	35.3	5.5	12.9	0.13	1.65	0.65	1.3
significance [a]	ns	***	ns	**	ns	ns	ns

[a] level of significance: **, *** $P < 0.01$, $P < 0.001$; ns = not significant.

Instead of a broadcast, while surface covering establishment of forage legumes, not more than one row of a preferably non-climbing legume between two cassava rows could be used to minimize competition and soil disturbance. This planting pattern was studied in Quilichao with 14 forage legumes. Differences in cassava yields were not significant and the competition appeared to be slight. Root yields ranged from 90 % to 110 % of sole cropped cassava in the first year. Strong yield reductions (40 to 55 %) were reported, when cassava was intercropped with already well established forage legumes in Quilichao (CIAT, 1994). Precipitation was low in the beginning of the cassava cropping cycle. About half of the yield decrease was attributed to competition by the legumes and half of it to minimum tillage. In West Africa, legumes as live mulch in maize competed for soil nitrogen to the disadvantage of maize in newly established plots (Mulongoy and Akobundu, 1985). In their trial legumes were not cut and removed. However after three years of continuous intercropping, legumes contributed nitrogen to the maize crop.

4.3 *Conclusions*

Rainfall characteristics (amount, erosivity, distribution), soil cover and soil surface conditions at the time of rainfall are the most important factors governing the erosion process. One heavy rainstorm at the beginning of the first cassava cropping season in Quilichao caused 70 to 80 % of the total annual soil losses in traditional flat cultivated and intercropped cassava. Establishment of forage legumes was difficult. In Quilichao, soil losses increased with increasing degree of "human traffic" for establishment of cassava and forage legumes. They were greatest in cassava associated with *Z.glabra* and exceeded losses from four-year old permanent bare fallow plots by 44 % in the first USLE crop stage period. In Mondomo, highly erosive storms were delayed until October 1990. Soil did not seem to be as susceptible to soil disturbance caused by establishment operations as in Quilichao. Losses were greatest in flat cropped cassava and reached the upper limit of tolerable soil losses, suggested by Reining (1992). Rain of low erosivity fell on both locations in the first three months of the second cropping period and caused very little soil loss. *Z.glabra* and *C.acutifolium* in Quilichao, *Z.glabra* and *C.macrocarpum* in Mondomo formed a dense, uniformly distributed soil cover and prevented erosion by the time more erosive rain fell. Soil losses were large in Quilichao, where forage legumes were not uniformly distributed. The same tendency could be seen in Mondomo. Dense and gapless stands of contour grass barriers effectively controlled erosion. *V.zizanioides* established slowly, but then formed a dense, very promising soil conservation barrier. The production of 2 - 3 t dry matter of vetiver grass per ha is remarkable on these small contour strips. The material is highly resistant to decomposition and might be used as mulch material or for other farm purposes.

Precipitation was far below average in Mondomo. This largely explains smaller soil losses than in 'normal' years when losses even on gentle slopes may easily exceed tolerable soil losses.

When enough water was available to guarantee a good emergence of cassava on the ridges, contour and traditional flat cropping achieved highest root yields. Simultaneously established forage legumes decreased cassava yields in the first cropping period by 10 to 20 %. Competition by legumes and probably negative effects of minimum tillage reduced yields in the second period by 30 to 40 %; *Z.glabra* possesses a more superficial and dense root system (Muhr et al., 1995), seeds freely and competes more with cassava than the other tested legumes. During the unusually dry beginning of the second cropping period, competition seemed to be mainly for water.

Contour barriers of grasses such as *P.purpureum* reduce area for cassava cultivation and depress root yields by competition, but produce valuable fodder material.

Cropping practices, recommended for the study region, to control erosion and achieve acceptable yields, are planting of cassava on properly constructed contour ridges where slopes are not greater than 15-20 %, and the use of contour barriers of densely planted grasses on slopes greater than 5 %. Another possibility is the use of minimum tillage on less eroded sites with better structured soils. The effectiveness of well established legumes such as *C.acutifolium*, *C.macrocarpum* or *Z.glabra* as contourstrips at adequate distances may also be concluded from the results of the present study. Roose (cit. by Fournier, 1967) studied the effect of buffer strips in Ivory Coast and found that the inclusion of taprooted legumes (Pueraria or Flemingia) in grass barriers established within cassava fields improved the absorption capacity of the strips and hence reduced runoff and soil loss compared to sole grass strips.

A combination of different control practices will be necessary to diminish soil erosion to acceptable levels on slopes steeper than 20 %.

Part 5 *LOSSES OF ORGANIC MATTER AND NUTRIENTS BY WATER EROSION*

In many regions of the humid and subhumid tropics, shallow topsoils with reasonably high organic matter and nutrient concentrations overlie infertile, acid subsoils. Soil losses on sloping agricultural land may easily exceed the tolerance limits, above which a sustained production is impossible. Erosion decreases available soil moisture capacity, rooting depth, organic matter content and nutrient availability and consequently productivity. Chemical soil fertility may be partly restored by mineral fertilization, whereas loss of organic matter and degradation of physical soil properties are often irreversible. In the tropical Andes, smallholders rarely fertilize their crops, especially cassava, which is grown on heavily eroded hillsides. Soil losses as high as 100 t ha^{-1} per cropping period are reported for cassava production systems in the South Colombian Andes on moderate to steep slopes (Howeler, 1985a). This means a loss of one centimeter of fertile topsoil per year. Soluble nutrients in runoff water contribute to the eutrophication of lakes and rivers, sediments fill up water reservoires and diminish the lifetime of electric power stations.
Erosion is a selective process: organic matter and nutrient contents of sediments are often higher than those of the matrix soil. Nutrients and chemicals are carried primarly by organic matter and the finer soil particles due to their relativly greater specific surfaces (Young et al., 1986). Smaller and less dense soil particles are preferably transported, coarser and denser particles more readily deposited. Rill flow transports a greater proportion of larger particles as compared with interrill flow, due to basic differences in the transport mechanisms (Alberts et al., 1980). Deposition, rather than detachment by raindrop impact, seems primarily responsible for changes in the textural and chemical composition of the sediment (Foster et al., 1985). Sediments eroded from agricultural soils consist of both soil aggregates and primary particles (sand, silt, and clay). Aggregate size and percentage of aggregated sediment depend on clay content, clay mineralogy, organic matter content , and iron/aluminium oxide contents (Meyer et al., 1992). When runoff water separates soil particles into those too large and heavy for removal and those transported with flowing water, an enrichment of transported sediments with fine particles and thus organic matter and plant nutrients may occur. The ratio of organic matter and nutrients present in the transported sediments over organic matter and nutrients in the matrix soil is called the enrichment ratio. Heavy textured soils produce highly aggregated sediment with much lower enrichment ratios than sediments of sandy soils, which are

greatly enriched in fine particles, not easily deposited (Knisel and Foster, 1981). Enrichment ratios are generally inversely related to the amount of soil loss (Massey and Jackson, 1952). The selectivity of the erosion process is more pronounced for those hydrologic events resulting in lower soil losses. Reductions in soil loss accomplished by conservation tillage or cropping management practices do not necessarily bring about comparable reductions in losses of crop nutrients (Young et al., 1986). Enrichment ratios, reported by Barrows and Kilmer (1963) for the USA, range from 1.2 to 4.7 for organic matter, 1.1 to 5.0 for total N, 1.3 to 3.3 for P, 1.4 to 12.6 for K, 1.0 to 2.4 for Ca, and 1.4 for Mg. Ratios are generally higher for the available fraction of nutrients (Rogers, 1941, cited in Barrows and Kilmer, 1963; Sharpley, 1985)). Tillage and antecedent moisture conditions and residue cover (Alberts and Moldenhauer, 1981), leachates of living plants or residues (Barisas et al., 1978; Alberts et al., 1978), slope and rainfall intensity (Lal, 1976a; Sharpley, 1980) may affect enrichment ratios.
Lal (1976a) working with Alfisols in West Africa found high ratios for C (2.4) and plant available P (5.8). Ratios for N were 1.6, for K 1.7, for Ca 1.5 and for Mg 1.2. In the South Colombian Andes, Reining (1992) reported no or only marginal nutrient enrichments for organic matter and total N; sediments were enriched by the factor 1.5 for Mg and 1.4 for K.
Available phosphorus and potassium were found to be predominantly associated with sediments (Munn et al., 1973; Burwell et al., 1975; Reddy et al., 1978). Additionally, Maass et al. (1988) reported higher losses of nitrogen, calcium, magnesium and sodium in the sediments than in the runoff. Losses of soluble nutrients in runoff water were generally low. Reining (1992) found a higher proportion of nutrient loss in runoff water, when soil losses were low.

5.1 *Losses of organic matter and nutrients in sediments and water runoff*

5.1.1 Soil loss and water runoff

Soil losses in Table 16 correspond to losses from original slopes with gradients between 7 and 13 % in Quilichao and between 12 to 20 % in Mondomo (uncorrected by USLE S- and L-factors). They were always significantly lower on cropped plots than on continuous bare fallow pplots. Within cassava cropping systems losses were greater in Quilichao than in

The interactions "cassava cropping systems x cropping period" and "cassava cropping systems x location x cropping period" were significant at levels of $P<0.001$ and $P<0.01$, respectively. Soil losses in Quilichao in the first cropping period were greatest, when cassava was intercropped with forage legumes. Practices to establish the legumes and highly erosive rainfall shortly after legume sowing made the soil very susceptible to erosion. In Mondomo, the soil seemed to be more tolerant of human traffic causing soil disturbance. Furthermore, erosive rain events were delayed until October 1990 and soil losses of the intercropping treatments were lower than when cassava was solecropped. Differences between cassava cropping systems were not significant in Mondomo.

Legumes, once well established, effectively controlled soil losses. The erosivity of precipitation and consequently the losses from bare fallow plots were greater in the second cropping period in Quilichao, but most erosive rain fell from the end of November on, when soil cover of cassava, legumes, weeds and mulch was high enough to prevent excessive losses from cropped plots.

Rainfall in Mondomo was less erosive in the second period. Soil losses were significantly greater in sole cassava and in association with *C.acutifolium* than when interplanted with contour grass barriers.

Table 16. Soil losses from permanent bare fallow plots and different cassava cropping systems in the cropping periods 1990-91 and 1991-92 in Quilichao and Mondomo.

	QUILICHAO		**MONDOMO**	
Treatments	1990-91	1991-92	1990-91	1991-92
	$t\ ha^{-1}$			
Bare fallow	143.7	222.6	227.9	160.7
Contour ridges	3.0 bc [a]	4.0 a	2.0 a	1.7 a
Flat cropping	8.2 bc	4.9 a	12.8 a	1.7 a
+Z.glabra	28.1 a	1.0 a	6.8 a	0.6 ab
+C.acutifolium	13.1 bc	1.6 a	2.6 a	1.7 a
+V.zizanioides	1.7 c	1.6 a	4.8 a	0.5 b
+P.purpureum	4.1 bc	4.1 a	1.1 a	0.4 b
+P.phaseoloides	15.4 ab	5.1 a	-	-
+C.macrocarpum	-	-	5.6 a	0.8 ab

[a] Analysis of variance for the soil losses from cassava cropping systems, common on both locations, are shown in Table 17. Values in a column with the same letter are not statistically different from each other.

Generally the best practices to control soil losses and to produce acceptable yields were the cropping of cassava on contour ridges and with contour grass barriers of dense stands.
The proportion of rainfall, lost as runoff, was generally low. Runoff in both cropping periods was greatest in the bare fallow treatment. Differences between cassava treatments, locations and interactions were not significant in the first year. In the second cropping period (Fig. 24) more water ran off in Quilichao (105 mm) than in Mondomo (45 mm; $P < 0.01$).

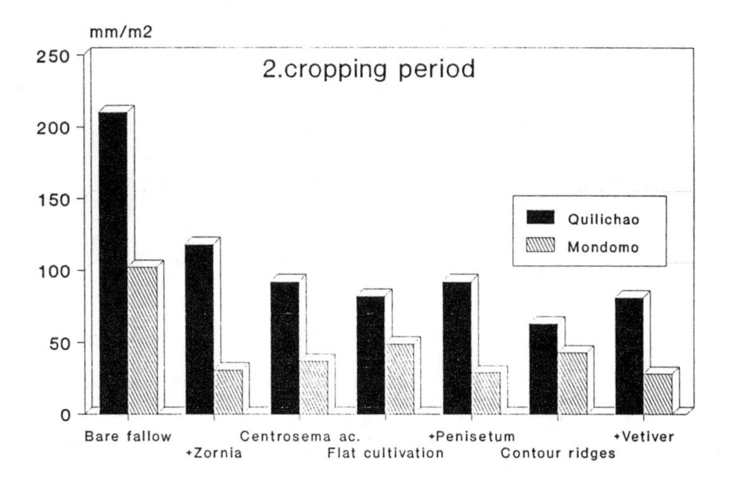

Figure 24: Runoff losses (mm m^{-2}) from bare fallow and different cassava based cropping systems in the second cropping period in Quilichao and Mondomo.

5.1.2 Organic matter and nutrient losses

Losses of organic matter and nutrients followed the same pattern as soil losses (Table 17). First of all they depended on the fertility status of the soil, the crop grown, and management practices used. They were a function of the amount of lost soil and the nutrient concentration of that soil.

In Quilichao, losses of Bray-II P from the cassava-*Z.glabra* association and of exchangeable Mg from all intercropping treatments were greater than those from the bare plots. Losses of the other nutrients and organic matter were significantly smaller. In Mondomo, always less organic matter and nutrients were removed from cropped than from continuously clean-tilled fallow plots.

Table 17. Losses of organic matter and macronutrients, and analysis of variance for nutrient losses from different cassava cropping systems in Quilichao and Mondomo during the cropping periods 1990-91 and 1991-92.

	bare fallow		**flat cropping**		**inter-[a] cropping**		**grass [b] barriers**	
	1990-1991	1991-1992	1990-1991	1991-1992	1990-1991	1991-1992	1990-1991	1991-1992
	kg ha^{-1}							
QUILICHAO								
OM	8314	12584	495	331	1284	94	203	212
total N	257	345	19.5	9.5	50	3.0	6.8	6.6
Bray-II P	0.59	1.11	0.17	0.18	0.49	0.03	0.05	0.08
exch. K	4.34	5.44	0.72	0.50	2.02	0.11	0.23	0.26
exch. Mg	1.10	1.35	0.64	0.52	1.61	0.14	0.28	0.31
MONDOMO								
OM	11302	7503	783	110	321	77	206	30
total N	441	272	33.9	3.9	12.7	7.9	7.9	1.1
Bray-II P	0.72	0.51	0.17	0.02	0.07	0.02	0.03	0.01
exch. K	6.24	3.95	0.95	0.17	0.45	0.13	0.30	0.05
exch. Mg	1.94	1.01	0.87	0.14	0.39	0.10	0.26	0.04

Analysis of variance

	soil loss	organic matter	total N	Bray-II P	exch. K	exch. Mg
Treatment (TRT)	*	*	*	ns	**	*
Location (LOC)	*	*	ns	*	ns	*
Cropping period (CP)	***	***	***	**	***	***
TRT x LOC	ns	ns	ns	ns	*	ns
TRT x CP	***	***	**	**	***	***
LOC x CP	ns	ns	ns	ns	ns	ns
TRT x LOC x CP	**	**	**	**	***	**

[a] Cassava intercropped with *Z.glabra* and *C.acutifolium*.
[b] Cassava with contour barriers of *V.zizanioides* and *P.purpureum*.

Soil analyses showed a sharp decrease in soil fertility on bare plots, which can not be explained by physical removal of topsoil only, especially for exchangeable Ca, Mg and K (Table 2). For that reason higher concentrations of organic matter and nutrients in sediments from cropped, fertilized plots led to relatively greater losses than from clean tilled fallow plots. At Quilichao in the first cropping period, cassava planted on the flat, lost about 5.7 % of the amount of soil lost from the bare plots, but 6.0 % of the organic matter, 7.6 % of total N, 58.2 % of exchangeable Mg, 16.6 % of exchangeable K and 28.8 % of Bray-II P. When cassava was associated with *Z.glabra*, soil loss was 19.6 % of that lost from the bare fallow treatment, but 189.7 % of its exchangeable Mg. Cropped plots received fertilizer annually, which in particular improved the levels of Bray-II P compared to the original soil, explaining the low P losses of the cassava-*V.zizanioides* treatment. It was established on old grassland of very low available P-status (2-4 ppm) four and three years after the other treatments in Quilichao and Mondomo, respectively.

In the second cropping period highly erosive rain fell when crop cover was well established. Removal of soil, organic matter and nutrients from the fallow plots in all cases was significantly greater than from plots with different cassava cropping systems; as for soil losses between cropping systems no significant differences occurred in Quilichao. In contrast, differences between cropping systems occurred in Mondomo, where elephant grass barriers reduced soil, organic matter, and nutrient losses most significantly. This treatment has been among the least effective in Quilichao. On the other hand, cassava associated with *C.acutifolium* lost most soil and nutrients in Mondomo, but was among the most effective soil protection treatments at Quilichao, together with the cassava-*Z.glabra* association and with vetiver grass barriers. Reasons for these interactions were the degree of soil cover and the quality of the stand of the contour grass, which showed an opposite behaviour at the two locations.

The differences between losses of organic matter and total N between the second and the first cropping period followed the differences in soil losses. No significant differences were found between treatments, when the bare fallow plots were included in the statistical analysis. But when analysis was performed on cassava cropping systems only, significant differences among individual treatments became evident. Cassava associated with *Z.glabra* reduced soil losses most, followed by cassava intercropped with *C.acutifolium* and cassava solecropped on the flat. Interactions of "locations x treatments x cropping periods" were all highly significant ($P < 0.001$). In Mondomo, all treatments lost less soil, organic matter and nutrients in the second year with insignificant differences between the cassava cropping systems. In the second year, higher erosivity (30 % more than 1990-91) in Quilichao increased losses from bare fallow plots, but with the exception of cassava on ridges their late occurrence in the cropping season led to lower losses on cropped plots. Compared to the previous cropping period, losses of soil and organic matter were largely reduced in intercropping systems. Soil losses were reduced by 27.1 and 10.4 t ha^{-1}, and losses of organic matter by 1,666 and 730 kg ha^{-1}, when cassava was intercropped with *Z.glabra* and *C.acutifolium*

Runoff was analysed for soluble P, K and Mg in the second cropping period. P losses were greater on the bare plots ($P < 0.01$) and greater in Quilichao than in Mondomo ($P < 0.01$); differences in amounts of K and Mg, lost in runoff, between treatments and locations were not significant.

Sediment-bound losses are generally greater than nutrients lost in runoff (Alberts and Moldenhauer, 1981; Maass et al., 1988). The proportion of nutrients lost in runoff increases, when soil loss is low or reduced by erosion control measures (Reining, 1992), because soil loss is reduced more than runoff. Fig. 25 shows the percent contribution of sediment-bound Bray-II P, exchangeable K and Mg to the total loss of each of these nutrients in the second cropping period. Between 70 and 85 % of total nutrient losses occurred with sediments from bare plots, which is a much higher proportion as compared to that observed in the cropped plots. Taking into account some overflow of runoff from this treatment in heavy rainstorms, the contribution of nutrients lost with water to total losses would even have been greater. The high proportion of nutrient losses by surface water stresses the importance of runoff control along with attempts to control soil losses to diminish nutrient losses and water contamination. In Quilichao, cassava intercropped with *Z.glabra* lost more water than any other cropping system, but least soil: Nutrients lost by water were 74 % of total P losses, 95 % of total K losses and 84 % of total Mg losses. Except for the fallow plots, the proportion of sediment bound K losses were always lower than those of P and Mg lost in sediments.

Perhaps interflow water coming up to the soil surface at the lower end of the plot, enriched in nutrients by leaching, might have contributed to this high nutrient loss by water (Barnett et al., 1971). Lal (1976c) found higher concentrations of N (two to threefold) and Mg and Ca, but not of K, in interflow water, compared to surface runoff. P concentrations of interflow were negligible.

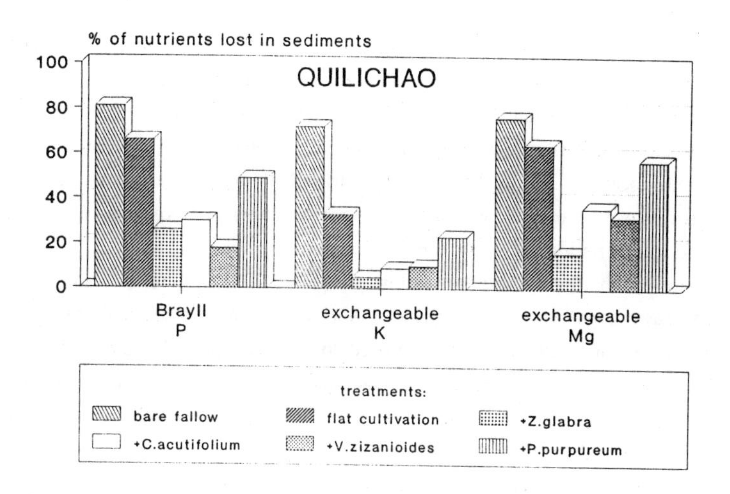

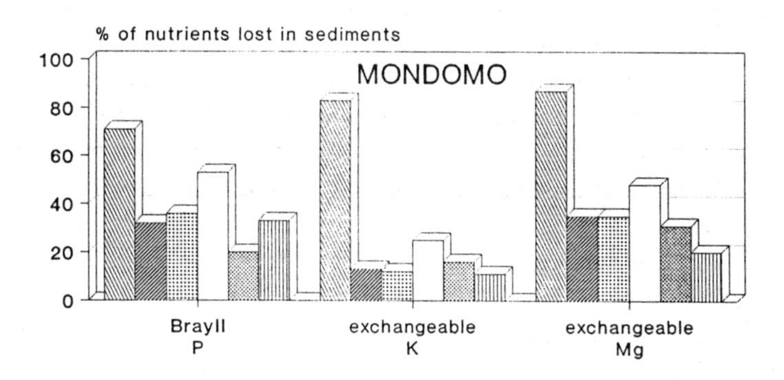

Figure 25:. Sediment bound nutrient losses as percentages of total losses in the second cropping period in Quilichao and Mondomo

5.2 Concentrations of organic matter and nutrients in sediments and water runoff

5.2.1 Concentrations in sediments

Nutrient concentrations in sediments are given in Table 18. Differences between the cassava cropping treatments were not significant with the exception of Bray-II P.
The concentrations in sediments were closely correlated with concentrations of corresponding matrix soils ($r = 0.79$ to 0.97) for organic matter, total N, exchangeable Mg and K on both locations and additionally for Bray-II P at Quilichao. At Mondomo correlations for P were clearly lower in both years. Among micronutrients, which were analysed in the second period, only Mn correlated well.
High erosion losses from permanent bare fallow plots since 1986 in Quilichao and 1987 in Mondomo decreased concentrations of organic matter, total N and exchangeable Mg, K and Bray-II P in source soils and sediments compared to cropped, annually fertilized cassava plots (Table 16 and Table 18). Sediments from plots with cassava and vetiver grass reflected the higher organic matter and lower P contents of original, newly cleared grassland soils of the study region.
With the exception of the concentration of exchangeable K, no significant interactions between treatments, locations and cropping periods occurred for organic matter and macro nutrients (Table 18).
Contents of total N were higher in Mondomo than in Quilichao and decreased in the second period. Mg contents, which were lower in Mondomo, increased in the second cropping period. Increase in the Mg concentration was slightly higher, when cassava was intercropped with forage legumes or with contour barriers of *P.purpureum*; this may indicate a possible influence of reduced tillage, combined with a lower degree of dolomitic lime incorporation in the intercropping treatments, although differences between treatments were not significant.
K concentrations in soils were higher in Mondomo with 0.20 $cmol \cdot kg^{-1}$ ($P < 0.01$; results not shown) than in Quilichao with 0.14 $cmol \cdot kg^{-1}$; no statistical differences were found between locations, when sediment concentrations were compared. This may indicate a higher selectivity for K in Quilichao. In Mondomo, however, contents in sediments increased in the second cropping period, whereas no change occurred in Quilichao, except when cassava was associated with forage legumes; the decrease in K contents is probably caused by the removal of K by the forage material.

As for exchangeable K, a possible higher selectivity in Quilichao was obtained for total and organic P, which were analysed in the second cropping period only.
Nutrient concentrations are generally found to be inversely related to the amount of eroded sediments (Massey and Jackson, 1952). No consistent trend was found for the sediments of the studied soils. There was a tendency for a positive correlation between total N content and soil loss.

Table 18. Concentrations of organic matter and nutrients in sediments from soils under permanent bare fallow, cropped with cassava (averaged over all cropping systems) and analysis of variance for cassava cropping systems.

	OM	total N	Bray-II P	exch. K	exch. Mg
QUILICHAO	%	mg kg^{-1}	mg kg^{-1}	-- cmol kg^{-1} --	
1990-91 cropping period					
bare fallow	5.8	1780	4.1	0.08	0.06
cropping systems	6.7	2360	19.2	0.23	0.73
1991-92 cropping period					
bare fallow	5.7	1565	5.2	0.06	0.05
cropping systems	7.1	2159	24.8	0.23	0.83
MONDOMO					
1990-91 cropping period					
bare fallow	5.0	1933	3.1	0.07	0.07
cropping systems	7.1	2732	14.2	0.24	0.63
1991-92 cropping period					
bare fallow	4.7	1697	3.2	0.06	0.05
cropping systems	6.8	2466	14.1	0.29	0.70
	Analysis of variance [a]				
Treatment (TRT)	ns	ns	*	ns	ns
Location (LOC)	ns	*	ns	ns	*
Cropping period (CP)	ns	***	ns	*	**
TRT x LOC	ns	ns	ns	ns	ns
TRT x CP	ns	ns	ns	ns	ns
LOC x CP	ns	ns	ns	*	ns
TRT x LOC x CP	ns	ns	ns	*	ns

[a] Analysis of variance without the bare fallow treatment; ns=not significant; *, **, *** significant at a level of $P<0.05$, 0.01 and 0.001, respectively.

5.2.2 Concentrations in runoff

Concentrations of soluble K and Mg in runoff water were lower from the bare plots than from cropped plots (Table 19). However, P concentrations were higher in runoff from the bare fallow plots in spite of lowest concentrations of P in their sediments. This might suggest, that runoff water from heavily eroded Andean Inceptisols contributes to a large extent to the contamination of streams and lakes. P concentrations above 0.03 mg kg^{-1} are considered high enough to stimulate the growth of algae and other aquatic plants in surface waters (Alberts et al., 1978). No significant differences between runoff nutrient concentrations among cassava cropping systems were found. The runoff concentrations of P in cassava intercropped with forage legumes and with barriers of *V.zizanioides* tended to be somewhat lower. On a single event basis runoff amount and nutrient contents were not well correlated, however, there was a weak positive correlation between runoff and P contents in Quilichao. It was highest, when cassava was cropped with elephant grass contour barriers. In Mondomo, the correlations were mostly negative, reaching significance in cassava intercropped with *C.macrocarpum*. Nutrient concentration of runoff showed a high variability. For two of three replicates of the bare fallow treatment in the first cropping period in Quilichao K and Mg contents of sediments and surface water were significantly correlated ($P < 0.05$). This was not the case with P. Similar results were found in Mondomo. In the second cropping period no significance was reached on both locations.

Table 19. Concentrations of soluble P, K and Mg in runoff water in the second cropping period 1991-92 (averaged over both locations).

Nutrient	bare fallow	cassava cropping systems	Significance [a]
	------- mg kg^{-1}	--------	
P	0.15	0.09	ns
K	0.92	1.51	ns
Mg	0.18	0.40	**

[a] Comparison of the means of concentrations in runoff water from bare fallow (control) with cassava cropping systems was done using the Dunnet-Test. Concentrations in runoff water among cassava cropping systems were not statistically different; ns = not significant, ** = significant at a level of $P < 0.01$.

The degree of fertilizer incorporation determines nutrient concentrations in runoff, as shown by Timmons et al. (1973) for N and P. In the second cropping period only planting holes for cassava in the legume sward were prepared. In these treatments dolomitic lime was incorporated on about 15 % of the plot area. No increase of K and Mg concentrations could be observed for these treatments compared to plots, where lime was incorporated completly. One reason may be the low annual lime rate of 500 $kg \cdot ha^{-1}$.
Leachates from plants and crop residues are sources of soluble nutrients (Tukey et al., 1958; Timmons et al., 1970) and are thought to increase concentrations in surface water. No indication was found, that leachates from forage legumes and their residues contributed to increased soluble P, K or Mg concentrations. There may exist a combined effect of leachates from cassava, legume, mulch and weeds: When cassava was intercropped with legumes, competition occurred; cassava development was slower and soil cover much smaller, compared to solecropped cassava. So leachates from solecropped cassava plants and their residues (rich in nutrients) may have offset increases in nutrient concentrations by forage legumes in the intercropping treatments.
Lal (1976b) observed highest concentrations of nutrients in runoff water immediately after fertilizer application. His finding could not be confirmed in the present and Reining's study (Reining, 1992).

5.3 *Selectivity of the erosion process*

Erosion is a selective process with regard to both chemical and physical soil properties. Smaller and less dense particles are more easily removed. Nutrients or other chemicals are primarily associated with organic matter and colloidal particles. Concentrations of organic matter and nutrients such as nitrogen decrease as soil losses increase (Massey et al., 1953; Stoltenberg and White, 1953; both cited in Barrows and Kilmer, 1967; McDowell and McGregor, 1984). Selectivity can be expressed by the enrichment ratio (ER), which is the ratio of the concentration of a component in sediments to that of the corresponding matrix soil.
Research by Lal (1985) on Alfisols in western Nigeria showed, that the rate of decline in the maize yields caused by natural erosion (2.5 cm soil lost) was over 16 times greater than by manual desurfacing to a similar depth. He found ER's of eroded sediments from these plots of 3.5 for organic matter, clay and nutrients. Selectivity depends on the degree of

Table 20. Enrichment ratios of organic matter, macronutrients and texture classes for sediments in the first and the second cropping period in Quilichao and Mondomo. [a]

	OM	total	----- exch. -----			Bray-II	soil texture		
		N	Ca [b]	Mg	K	P	Sand	Silt	Clay
Quilichao									
1990-91	0.93	0.99	1.39	1.04	1.30	1.07	1.20	0.89	0.98
1991-92	0.97	0.96	1.33	1.12	1.20	1.25	1.14	0.87	1.00
Mondomo									
1990-91	1.03	1.14	1.23	1.13	0.93	1.13	1.18	0.96	0.92
1991-92	1.03	1.03	1.27	1.03	1.10	0.98	1.13	0.88	0.99
Analysis of variance									
Treatment (TRT)	ns	ns		ns	ns	ns	ns	ns	ns
Location (LOC)	*	*		ns	ns	ns	ns	ns	ns
Cropping period (CP)	ns	ns		ns	ns	ns	*	***	**
TRTxLOC	ns	ns		ns	ns	ns	*	ns	ns
TRTxCP	ns	ns		ns	ns	ns	ns	***	ns
CPxLOC	ns	ns		ns	*	ns	ns	ns	ns
CPxLOCxTRT	ns	ns		ns	ns	ns	ns	ns	ns

[a] For calculating enrichment ratios of sediments, frames for bare fallow and cassava cropping systems were averaged, as no significant treatment effects were observed in both cropping periods.

[b] Ca enrichment ratios for sediments from continuously clean tilled fallow plots (uncontaminated samples).

aggregation of the sediment. The sediment load from sandy soils is greatly enriched with fine components; coarser particles are deposited already in the field and enrichment ratios are high. However, the Inceptisols of the study sites show an excellent aggregation: they are high in organic matter and clay content, contain sesquioxides, kaolinite and amorphous materials in the clay fraction. Sediments from those soils are generally highly aggregated as well. Clay and silt are much more evenly distributed across particle classes in highly aggregated sediment (Foster et al., 1985). Therefore, little selectivity will be observed for this condition.

ER's for sediments from plots in bare fallow and with cassava intercropped with forage legumes confirm this observation. With the exception of exchangeable Fe, ER's of sediments from bare fallow plots did not show significant differences to those from cropped plots. Therefore, only ER's for sediments averaged across these treatments are presented in Tables 20 and 21. Concentrations of organic matter and nutrients in sediments were compared with those of the first ten centimeters of topsoil. Below that depth concentrations decreased sharply, especially for Bray-II P, exchangeable Mg, Ca and K. In Quilichao, ER's of organic matter and total nitrogen were lower than in Mondomo ($P<0.05$) as sediment concentrations did not exceed those of matrix soils in Quilichao. The ER for exchangeable K was significantly higher in the first cropping period in Quilichao than in Mondomo. This higher selectivity for K by water erosion in Quilichao was confirmed by significantly lower concentrations in topsoils, but insignificant differences in sediment concentrations compared to Mondomo. ER's of Fe were higher for sediments from bare fallow (1.24) than from cassava-*C.acutifolium* (0.87). In Mondomo highest ER's of B were found in the cassava-*C.acutifolium* association and lowest in bare fallow. In Quilichao ER's of B were in the reverse order.

High ER's of 6.13 for Bray-II P and of 1.97 for exchangeable Mg in sediments from cassava with barriers of *V.zizanioides* were found in the first cropping period in Quilichao, these ratios being significantly higher than in the other cropping systems ($P<0.05$; data not shown). The plots of this treatment were newly established in June 1990 on old grassland, whereas the other plots were already cropped with cassava for four years in Quilichao and three years in Mondomo (old plots). Lal (1976a) reported that immediately after forest clearing enrichment ratios were highest and then decreased. Probably easily removable nutrients were lost in the preceeding years from old plots. Confirming this observation, Reining (1992) reported higher enrichment ratios of exchangeable Mg (1.3 to 1.7) and K (1.4) for sediments from the same plots, used in the present study, during 1987 to 1989. However, he compared sediment concentrations with concentrations of the 0 - 20 cm topsoil layer, not with those of the 0 - 10 cm layer, as done in this study.

Table 21. Enrichment ratios of exchangeable sulfur and micronutrients for sediments averaged over bare fallow and cassava cropping systems in the second cropping period at Quilichao and Mondomo.

	S	B	Cu	Zn [a]	Mn	Fe
Quilichao	1.36	1.19	1.07	1.44	0.85	1.18
Mondomo	1.50	1.01	0.80	1.22	1.41	0.84
		Analysis of variance				
TRT	ns	ns	ns		ns	*
LOC	ns	ns	**		*	ns
TRTxLOC	ns	*	ns		ns	ns

[a] Zn enrichment ratios for sediments from continuously clean tilled fallow plots (uncontaminated samples).

Sharpley (1985) found the enrichment of Bray-I and labile P significantly greater than for other P forms. Total and organic P were analysed in the second cropping period only. Sediments were highly enriched in organic P (ER's of 2.36 at Quilichao and 4.83 at Mondomo; data not shown), but not in total P (1.05 and 1.27 at the two sites, respectively) and Bray-II P.

Correlations between soil losses and enrichment ratios generally were low and not significant. The ER of exchangeable K seems to be inversely related to soil loss, whereas the ER of exchangeable Cu seems to increase with increasing soil losses. There exists a weak positive relationship between the enrichment ratios of organic matter and total N and organic matter and silt.

Sediments contained more sand than the source soil (Table 20). Less soil losses from cropped plots in the second cropping period probably led to a decline in sand enrichment, which was more pronounced for intercropped systems than for flat cropping. Reasons for sand enrichment may be, that clay and silt (together with organic matter and sesquioxides) are strongly aggregated, reaching sizes, not casily removable by erosive forces. Particle selectivity during the erosion process is associated with interrill erosion. Rill erosion is seldom limiting transport and, therefore, less selective (Foster and Meyer, 1975). It played an important role in the erosion process on bare fallow and on cassava plots shortly after their establishment in the first cropping period in Quilichao. Alberts et al. (1980) determined the particle composition of eroded aggregates. They found for all sizes > 0.05 mm (up to 2mm) aggregates to be enriched with sand. ER's of sand were higher in Mondomo than in Quilichao for sediments from bare fallow plots. Lal (1976a) observed increasing sand contents in sediments, when slopes increased and attributed the change in particle size distribution of eroded soil to the higher velocity of runoff water at steeper slopes and thus its higher capacity to transport coarser particles. Silt was not selectively removed by water erosion: Contents were lower in sediments than in matrix soils. ER's of silt for sediments from bare fallow decreased markedly (from 0.99 to 0.81) in the second cropping period compared to the other treatments.

Slope gradient, rainfall erosivity, soil water permeability and living or dead soil cover are all factors, which may affect runoff velocity and consequently sediment enrichment, and makes it difficult to distinguish between the influences of the individual factors in the present study.

5.4 *Conclusions*

Soil erosion by water reduces depth of fertile topsoil, where organic matter and most nutrients are concentrated. Loss of soil fertility by erosion depends on the amount of soil and runoff water lost and the concentrations of nutrients in these two components. Soil losses were highest on bare fallow plots. Among cassava cropping systems in Quilichao in the first cropping period cassava intercropped with forage legumes lost greatest amounts of soil.

The degree of human traffic for legume establishment operations determined the losses. They were in the order *Zornia glabra* > *Pueraria phaseoloides* > *Centrosema acutifolium*, followed by traditional flat cropping. Most effective treatments were cassava cropped on contour ridges or with contour barriers of grasses. At Mondomo, flat cropped cassava lost most soil. In the second cropping period soil losses were generally lower: Best erosion control was achieved when cassava was intercropped with *Z.glabra* and *C.acutifolium* and with vetiver grass in Quilichao, and with contour grass barriers in Mondomo.

Losses of organic matter and nutrients followed the same pattern as soil losses. Organic matter losses were high, according to its high contents in soils. These losses may be regarded as critical: Organic matter is one of the most valuable soil components and among the hardest to replace (Barrows and Kilmer, 1963). Losses of organic matter and nutrients were generally greatest from permanent bare fallow plots. Heavy erosion losses on the bare plots decreased soil fertility to low levels of available nutrients, especially Mg, Ca and K. Annual liming and fertilization of cassava maintained soil fertility at levels about equal to uncultivated soils and increased P levels four- to fivefold compared to uncultivated soils and bare plots. Therefore, organic matter and nutrient losses from cassava cropping systems were over-proportionally greater compared to fallow plots: In the first cropping period in Quilichao cassava associated with *Z.glabra* lost 19.6 % of the soil lost from the bare plot, but 63.8 % of exchangeable K, 131.2 % of Bray-II P and 189.7 % of exchangeable Mg with losses from bare fallow set to 100 %. Losses of nutrients with sediments from cassava cropping systems ranged from 26 to 1726 $kg.\cdot ha^{-1}$ for organic matter, 0.9 to 65.5 $kg \cdot ha^{-1}$ for total N, 0.03 to 2.08 $kg \cdot ha^{-1}$ for exchangeable Mg, 0.04 to 2.77 $kg \cdot ha^{-1}$ for exchangeable K and 0.004 to 0.77 $kg \cdot ha^{-1}$ for Bray-II P.

Concentrations of organic matter and macronutrients in sediments were well correlated with those of source soils. Somewhat lower correlations were found for Bray-II P in Mondomo. Among micronutrients only Mn correlated well.

On bare fallow plots, over 70 % of the total loss of available Mg, K and P was associated with sediments. When soil loss decreased, the proportion of runoff-bound soluble Mg, K and P increased. In cassava cropping systems, amount of K, lost by runoff water, always exceeded losses by sediments. In Quilichao in the second cropping period, over 90 % of K was lost with runoff water, when cassava was intercropped with *Z.glabra* and *C.acutifolium* and with grass barriers of *V.zizanioides*. Probably interflow water led to a enrichment of K and Mg in runoff water and increased the proportion of nutrients lost by runoff. Concentrations of K and Mg were higher in runoff water from cassava treatments than from bare fallow. P concentrations, however, were higher from the bare plots.
Enrichment ratios, defined as the ratio of the concentration of an element in the sediment to that in the source soil, were generally low, which is to be expected from sediments of highly aggregated soils. In Quilichao, ratios were found to be in the following order: Organic matter (0.96 - 0.97) and total N (0.93 - 0.99) < exchangeable Mg (1.04 - 1.12) < Bray-II P (1.07 - 1.25) < exchangeable K (1.20 - 1.30). There was a tendency for slightly higher enrichment ratios for organic matter and nitrogen in Mondomo and for exchangeable Mg, K and Bray-II P in Quilichao. Comparing the results of soil analysis of permanent bare fallow with annually fertilized cassava treatments, the greater selectivity of the erosion process for Mg, Ca and K is confirmed by higher enrichment ratios of those nutrients.
Sediments had higher sand contents than the original soils. The enrichment was more pronounced in Mondomo, probably caused by the steeper slopes. Clay and fine silt particles were highly aggregated and less removable than sand; this might explain the rather low enrichment ratios for organic matter and nutrients found in the present study.
Cassava is cropped mainly on shallow, acid soils of low fertility. It is often the last option for a farmer to grow a crop on exhausted soils. In the tropical Andes it is usually cropped by smallholders on moderate to steep slopes without any fertilizer. High soil losses and a selective impoverishment of nutrients, especially potassium, leads to very low cassava yields. Besides the prevention of high soil losses, erosion control measures have to take care of nutrient losses by runoff water also. Contrary to most results published on nutrient losses by water erosion in the literature, surface runoff in our study contributed substantially to total available nutrient losses.

Part 6 *IMPACT OF CROP MANAGEMENT AND WATER EROSION ON SOIL PROPERTIES*

Soil erosion reduces crop yields generally more in the tropics than in the temperate zones (Stocking and Peake, 1986). Decrease in yield is commonly attributed to a loss of nutrients and organic matter, a reduction in available water-holding capacity and topsoil depth and a decline of the soil's structural stability. Biological activity can also be reduced by erosion (Lal et al., 1979). Yield reduction by topsoil removal is both crop and site specific (Mbagwu et al. 1984). Fertilization can restore yields only under favorable climatic and soil physical conditions (Eck, 1968; 1969).
Organic matter and nutrients are lost with soil and surface runoff. Concentrations in sediments are generally higher than in the original soil, which leads to a selective impoverishment. Enrichment of nutrients in eroded sediment was found for exchangeable K, Mg and probably Ca on the locations of the present study.
The stability and size distribution of soil aggregates are fundamental characteristics of soil structure. They are among the most important physical properties governing resistance to erosion in many soils (Bryan, 1968). Water-stable aggregates are the basis of a good structure for the growth of plants; the sizes of aggregates control the pore-size distribution and consequently physical and chemical processes in the soil. Highly significant correlations were found between aggregate stability and organic matter in many soils, indicating that organic matter is mainly responsible for the stabilization of aggregates (Chaney and Swift, 1984; Kemper and Koch, 1966). Other authors report relatively low correlations (El-Swaify and Dangler, 1976). This may be due to a variety of reasons, among them those presented by Tisdall and Oades (1982). These authors argue that only part of the organic matter is responsible for water-stable aggregation; there is a content of organic carbon, above which there is no further increase in water-stable aggregation. Organic materials are not the major binding agents, thus it is the disposition rather than the type or amount of organic matter which is important.
Water permeability is another important characteristic of soil structure and depends on texture, clay mineralogy, air entrapment, aggregate stability and subsequent susceptibility to sealing and crust formation, as well as soil tillage and crop management. Sandy soils, highly weathered and well aggregated soils of the humid tropics or soils under long term bush fallow exhibit generally high infiltration rates (Lugo-Lopez et al., 1968; Wolf and Drosdoff, 1976; Lal, 1976a). High rates, measured by double-ring infiltrometers are

sometimes attributed to lateral flow, enhanced by biotic activity (Lal, 1979). Soils with weakly structured topsoils may loose rapidly their excellent water transmission properties after clearing (Wilkinson and Aina, 1976). Two years of clean tilled fallow decreased infiltration rates greatly due to heavy erosion losses (Lal, 1976d). Intercropping cassava increased infiltration compared to sole cropping on Alfisols in West Africa (Aina et al., 1979; Hulugalle and Ezumah, 1991). Improvements of soil chemical and physical properties after only two years of growing grasses and legumes on an eroded Alfisol, were reported by Lal et al. (1979): Organic matter, total N and cation exchange capacity, infiltration rate and soil moisture retention increased, particularly at low tensions, whereas bulk density decreased. Ameliorative ability depended on the species used. Intensities of heavy rainstorms in the tropics easily exceed even high infiltration rates. Soil crusts decrease rates by reducing the hydraulic conductivity of the top layer, even when underlying layers may be much more permeable, resulting in high water and often soil losses.

One of the most important yield-limiting effects of soil erosion is considered the decrease in available water-holding capacity (Frye et al. 1982; Langdale et al. 1979; Williams et al., 1981). It is low in the surface horizons of most upland soils in the tropics (Lal, 1987). Underlying layers may present physical and chemical limitations for crop production. B-horizons of Inceptisols of the study region are infertile, acid and highly Al-saturated; consequently effective rooting depth is restricted. There is only a limited number of crops adapted to those conditions, cassava being one of them. Connor et al. (1981) have shown, that when cultivated under water stress on soils of the study area, cassava is capable of extracting water from deeper soil layers. Thus, it is a crop suited for conditions where topsoil has been partially removed by erosion.

6.1 *Soil fertility*

6.1.1 Changes in chemical and textural properties over two cropping periods

Cropping systems averages of chemical and textural properties of soils and their statistical analysis (excluding bare fallow) are presented in Tables 22 and 23. Chemical properties of soils generally did not differ significantly among cropping systems in the beginning of the trial period of the present study in May 1990. Greatest effects were found for chemical properties between the beginning of the trial period in May 1990 and its end in April 1992.

Soil pH did not change in Quilichao with the exception of an increase in ridged soil, but increased in Mondomo in flat and intercropped cassava. Organic matter content increased in Quilichao from 1990 to 1992; a 16 % increase was observed in the association with *Z.glabra*, compared to 8 %, when cassava was flat cropped. Decrease of organic matter content in Mondomo was lowest with *Z.glabra* (-6 %). However differences between cropping systems were not significant. In Australia, Rixon (1966) found lower pH values for soils under clover pastures than under grass pastures when soils were analysed five years after pasture establishment. Increase in soil acidity was found to be associated with reduced base saturation and increased organic matter content (Williams and Donald, 1957).

Table 22. Soil reaction, concentrations of organic matter and macronutrients of topsoils (0-20 cm) in May, 1990 and April, 1992 averaged over all cropping treatments, excluding bare fallow. Plots were cropped since 1986 in Quilichao and since 1987 in Mondomo.

	pH (1:1)	OM	total N	mineral P	Bray-II P	Ca	Mg	K	Al
		%	-------	mg kg^{-1}	-------	---------	cmol kg^{-1}		---------
QUILICHAO									
1990	4.5	6.1	2210	50	14.2	1.44	0.53	0.17	3.24
1992	4.5	6.7	2072	41	18.3	1.61	0.61	0.18	2.72
MONDOMO									
1990	4.5	6.5	2060	69	8.8	1.30	0.46	0.22	1.92
1992	4.6	5.9	2189	69	14.3	1.42	0.60	0.22	1.47
			<u>*Analysis of variance*</u>						
TRT	ns	ns	ns	ns	ns	ns	ns	ns	ns
LOC	ns	ns	ns	ns	ns	ns	ns	**	ns
CP	**	ns	ns	*	***	***	***	ns	***
TRTxLOC	*	ns	ns	ns	ns	ns	ns	ns	ns
TRTxCP	ns	ns	**	*	ns	ns	**	ns	ns
LOCxCP	ns	***	***	ns	ns	*	*	ns	ns
TRTxLOCxCP	*	ns	ns	ns	ns	ns	ns	ns	ns

Contrary to organic matter, the content of total N decreased in Quilichao and increased in Mondomo. Concentrations in soils, where cassava was intercropped with legumes, were lowest in 1990, but among the highest two years later (Fig. 26).

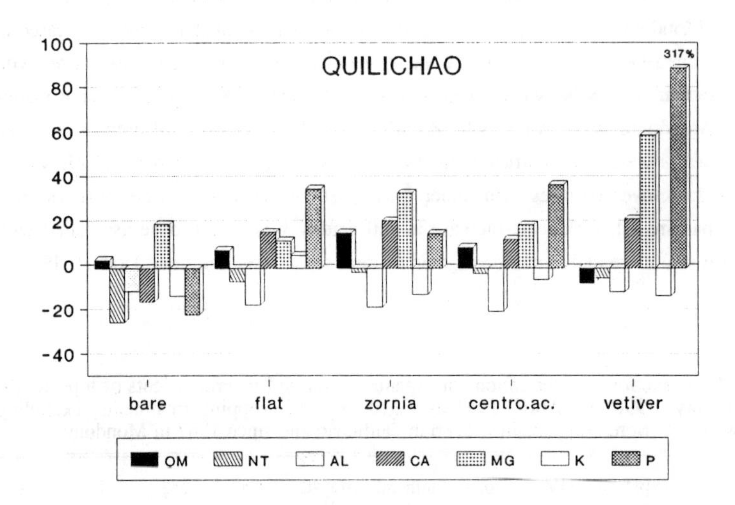

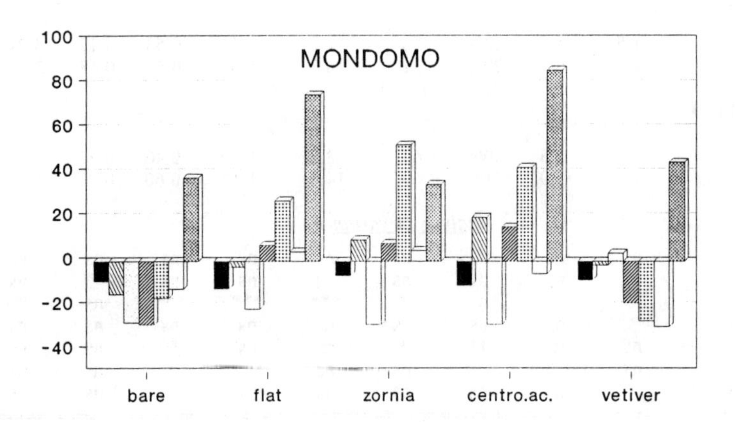

Figure 26: Changes in soil reaction and organic matter content of topsoils (0-20 cm) of bare fallow and cassava cropping systems in Quilichao and Mondomo from May 1990 to April 1992.

(*, **, *** = the difference in concentration is significant at a level of $P < 0.05, 0.01, 0.001$).

Concentrations of available P, Ca and Mg increased due to the annual fertilization and concentration of free Al decreased; the increase of Mg was more pronounced in Mondomo than in Quilichao. Concentrations were 36 % higher in intercropped cassava after two years, compared to a 19 % increase in flat cultivated, solecropped cassava. K concentrations did not change significantly from 1990 to 1992. They were higher in Mondomo than in Quilichao, probably due to less removal of K by roots under the low yielding regimes of that location.
Similar to the macronutrients, the time of cultivation had the greatest effect on the concentrations of micronutrients (Table 23). In 1992 increases of Zn concentrations were highest in ridged soils in Mondomo; they increased by 300 % compared to 53 % in Quilichao. Mn concentrations remained at the same level in Quilichao. In Mondomo, Mn concentrations were highly variable, in some cases reaching toxic levels for plant growth. Similar to Zn, increases in Mn were greatest in ridged soil (+128 %). Among cropping systems no significant differences in Fe concentrations were found in Quilichao; in Mondomo concentrations were significantly greater in soils of the cassava-*P.purpureum* treatment than when intercropped with *Z.glabra*.
Lack of significant differences between treatments is to be expected for the short period of two years of the present investigation. Forage legumes were cut regularly, stems and leaves were removed from the plots and not left as surface mulch, which would have probably caused greater changes in soil properties. Considerable amounts of nutrients were removed with the legumes. The idea behind the cut and carry system was, that smallholders in the cassava growing region of the south colombian Andes generally refuse to grow crops or fodder plants, if no direct benefit can be derived from such a practice. Additional nutrient loss on intercropped plots by the removal of the fodder plants was probably compensated by the lower extraction of nutrients with cassava, which suffered from strong competition by the legumes. Cassava biomass yields were greatly reduced. Lal et al. (1979) found significant improvements of soil chemical and physical properties after only two years when growing legumes and grasses on an eroded Alfisol. Legumes and grasses were left growing on the plots without cutting. In this way, organic matter, total N and cation exchange capacity, infiltration rate and soil moisture retention at low tensions increased whilst bulk density decreased.Usually short term changes in soil properties are confined to the first few centimeters of the topsoil. In this study, however, soil properties of the first 10 or 20 cm are compared so that any significant differences caused by management practices, will be diluted by the bulk of little or non-manipulated soil.

Table 23. Concentrations of micronutrients and texture of topsoils (0-20 cm) under averaged over different cassava cropping systems excluding bare fallow. Samples were taken in May 1990 and April 1992 from plots cropped since 1986 (Quilichao)and 1987 (Mondomo).

	S	B	Cu	Zn	Mn	Fe	Sand	Silt	Clay
	--------------------			mg kg^{-1}		------------------	--------	%	--------
Quilichao									
1990	64.0	0.29	0.77	0.74	9.3	30.0	19.8	19.3	60.9
1992	55.6	0.36	0.63	1.11	9.4	23.7	25.6	20.5	53.9
Mondomo									
1990	63.6	0.31	1.01	0.71	52.1	37.0	24.4	20.2	55.4
1992	44.6	0.31	0.94	1.45	88.5	25.1	26.7	22.5	50.8
				Analysis of variance					
TRT	ns	ns	ns	ns	ns	*	ns	ns	ns
LOC	ns	ns	ns	ns	ns	ns	ns	*	*
CP	***	**	***	***	***	***	***	***	***
TRTxLOC	ns	ns	ns	ns	ns	**	ns	ns	*
TRTxCP	ns	ns	ns	*	*	ns	ns	ns	ns
LOCxCP	ns	ns	ns	ns	*	ns	*	ns	ns
TRTxLOCxCP	ns	ns	ns	*	**	ns	ns	*	ns

6.1.2 Long-term changes

In order to assess changes in chemical soil fertility on the bare fallow plots since the beginning of the research project, soil samples were taken from adjacent grassland (Quilichao) or bush fallow (Mondomo), representing conditions present at the onset of the experiments in 1986 (Quilichao) and 1987 (Mondomo). Necessarily high soil losses and the absence of any vegetation degraded the plots in permanent bare fallow to a large extent. Without inputs of lime and fertilizer, agricultural production would no longer be possible. Accumulated soil losses were approximately four cm of topsoil from bare plots in Quilichao and over five cm in Mondomo until the beginning of the present study in May 1990 (Table 24). A sharp decline in pH, organic matter and concentrations of available nutrients was observed, especially for Mg, Ca and K. With the exception of total N, which still decreased markedly from 1990 to 1992, concentrations did not change significantly with increasing losses of soil in this time period. Probably concentrations of organic matter and nutrients are reaching a constant value after heavy erosion losses on bare plots, bringing about a chemical stability at a low level, at least for a certain time.

Table 24. Accumulated dry soil losses from permanent bare fallow and soil chemical properties in the top ten centimeters of bare fallow soils in May 1990 and April 1992 and reference data from grass-bush fallow representing conditions in 1986-87.

	QUILICHAO			MONDOMO		
	1986 [a]	1990	1992	1987 [a]	1990	1992
Accumulated soil losses ($t\ ha^{-1}$)	0	426 [b]	792	0	565 [b]	953
pH (H_2O)	4.6	4.1	4.0	4.7	4.3	4.4
OM (%)	7.9	5.8	6.2	7.7	5.0	4.6
Total N ($mg\ kg^{-1}$)	2352	2165	1643	2688	1960	1624
Exchangeable Ca ($cmol\ kg^{-1}$)	1.04	0.20	0.18	1.20	0.27	0.21
Exchangeable Mg ($cmol\ kg^{-1}$)	0.41	0.05	0.06	0.62	0.06	0.05
Exchangeable K ($cmol\ kg^{-1}$)	0.19	0.08	0.08	0.32	0.09	0.07
Exchangeable Al ($cmol\ kg^{-1}$)	3.95	5.43	4.63	2.70	3.00	1.96

[a] Analysis of original grassland soils in Quilichao and bush fallow soils in Mondomo in direct vicinity of the bare fallow plots in May 1990. Their concentrations are used as a reference to indicate the decline in soil fertility of continuously clean tilled plots by water erosion and tillage.
[b] Reining, 1992. Cadavid, CIAT, Colombia; pers. communication.

Al-saturation and free Al concentrations increased in the first years of bare fallowing. The horizon, which directly underlies the topsoil, however, has a lower Al concentration (REINING, 1992). With increasing soil loss, tillage operations mixed top and subsoil, which led to a decrease in Al in 1992.

Erosion is a selective process: Sediments are generally enriched in organic matter and nutrients compared to the soil, where they erode from. Enrichment ratios were low on the

study sites. The highest ratios were found for exchangeable K, Mg and probably Ca. Concentrations of organic matter were not increased in sediments.
This explains only partly the very low nutrient levels of topsoils of permanent bare plots. Original grassland soils show much higher nutrient levels in the depth 10 to 20 cm: Ca was 330 %, Mg 400 % and K 50 % higher than in the top 10 centimeter of bare fallow soils (data not shown). High losses especially of Ca and Mg by leaching may be a further reason for nutrient impoverishment on bare fallow.

Sediments may bury crops downslope and lead to greatly reduced yields or to a complete crop failure. However, soils may also be enriched by sediments, resulting in a better crop growth and increased yields. The organic matter and nutrient concentrations of bare fallow soils (0-10 cm) were determined in October 1991 for four slope positions. Results are presented in Fig. 27 as 100 times the ratio of nutrient concentration in a plot section to nutrient concentration averaged over all sections. Differences of concentrations between positions were generally significant. Free Al concentration was greater and exchangeable Ca concentration smaller in Quilichao than in Mondomo. No interaction between location and position occurred.
From plot establishment in 1986-87 to October 1991, accumulated soil losses were 570 t ha^{-1} at Quilichao and 780 t ha^{-1} at Mondomo. Rill erosion contributed to a large extent to soil and nutrient losses from bare plots. Original uniform slopes developed a slightly concave topography with accumulated soil losses. As expected at the lower end of the plot concentrations were increased by deposition of sediments. At the upper part of the plot, concentrations were higher, because soil and nutrient removal by rill erosion did not play such an important role and soil losses were much smaller than in the middle of the plot. Once rills are formed, surface water is concentrated and erosive power increases downslope.
Concentrations of free Al decreased, as topsoil loss increased. Together with the physical stability, which soils maintain after even high erosion losses, this may indicate the potential of eroded hillsides for crop production, once soil fertility constraints are alleviated.

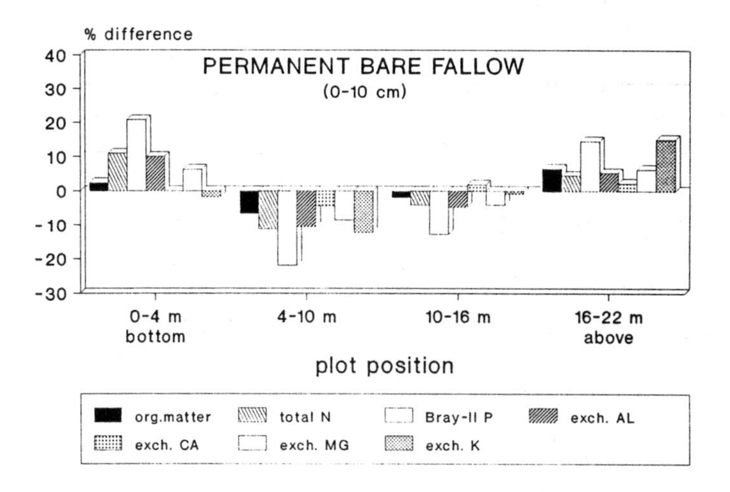

Figure 27: Deviation (%) in the concentrations of organic matter and nutrients from mean plot topsoil concentrations (0-10 cm depth) in different slope positions of bare fallow plots in October 1991, averaged over both locations.

6.2 *Changes in physical soil properties*

6.2.1 Aggregate stability

Fig. 28 shows meanweight diameters (MWD) of water-stable aggregates for both locations and both cropping periods. They were significantly greater in Quilichao than in Mondomo ($P<0.05$) in the first cropping period only. Interactions between locations and treatments were not significant, but between treatments significance occurred. Highest stabilities were found on plots, which were established 1990 on grassland in Quilichao and bush fallow in Mondomo (cassava-vetiver grass treatment), followed by contour ridged soils. Soils of the cassava-vetiver grass and contour ridge treatment showed highest proportions of water-stable aggregates > 2 mm and lowest proportions in the smaller

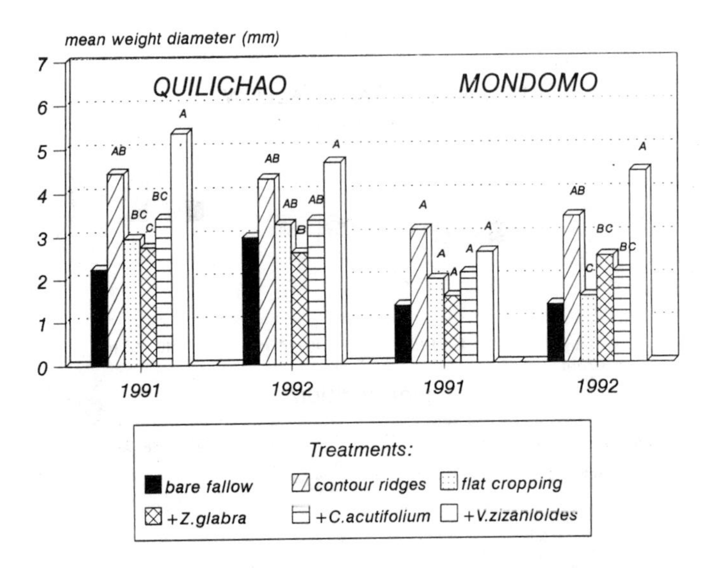

Figure 28: Meanweight diameter (mm) of water-stable soil aggregates of permanent bare fallow and cassava cropping systems in Quilichao and Mondomo at the end of the first and second cropping period.

aggregate sizes. No significant differences ocurred among cropping treatments and between cropping treatments and bare fallow soils for the proportion of microaggregates (< 0.25 mm), although there was a tendency to increase in flat, solecropped and intercropped cassava.

The soil stabilizing effect of grasses (mainly *Paspalum notatum*) is excellent and exceeds that of long-term bush fallow in Mondomo. These results confirm findings of Aina (1979) in western Nigeria, where percentage of water-stable aggregates (>2.36 mm) in the topsoil was greatest and rather constant under grass fallow compared to bush fallow. More intense drying and wetting of the soil of contour ridges compared to sole- and intercropped cassava on the flat probably increased MWD's to values similar to locations previously under grass. This extraordinary structure may be one important factor for high yields on ridges (Reining, 1992). The establishment of forage legumes in May 1990 was difficult, especially *Z.glabra*, which had to be reseeded three times. Trampling caused aggregate disintegration and lowest

aggregate stability among cropping systems in the first year. In Quilichao this effect seems to have endured until the end of the study period in 1992. Plots in permanent bare fallow were frequently tilled to break surface crusts and prevent the formation of deep rills. Tillage, the absence of crops, weeds and mulch, and heavy erosion losses had led to a decrease in the organic matter (Table 24) and, thus, decreased aggregate stability. The proportion of aggregates > 2 mm was lowest, compared to cropped treatments. Bare fallow is the most deleterious management treatment with respect to soil structure (Oades, 1984). Roots and fungal hyphae, particularly VA myccorrhiza, are associated especially with macroaggregates > 2 mm, as discussed by Tisdall and Oades (1982). They are considered as temporary binding agents, because generally they do not persist for more than one year.
Correlations between the stability of aggregates and the organic matter content of the topsoils were low ($r = 0.41$; $P<0.05$). Reasons are, that the contents of organic matter in the bulk of soil may differ from contents in dry aggregates of about 10 mm diameter, used for the wet sieving procedure, although for highly aggregated soils such as the Inceptisols in this study, differences were small. Secondly, organic matter may not be the only aggregate binding agent on these heavy textured Inceptisols; they contain considerable amounts of sesquioxides and amorphous materials (Reining, 1992), which may play an important role in soil aggregation. Furthermore only an "active" portion of organic matter and /or other soil components may play an important role in aggregate stabilization. Some authors consider the existence of a threshold value, above which no important increase in aggregate stability can be expected (Tisdall and Oades, 1982; Luk, 1979). This may have been the case in our plots where organic matter concentrations were high throughout. Even on long-term bare fallow plots concentrations in the top 20 cm never fell below four per cent.
A higher correlation was found for the relationships between aggregate stability and cation exchange capacity ($r = 0.54$; $P<0.001$), and exchangeable Al ($r = 0.45$; $P<0.01$). Exchangeable Fe did not correlate significantly with stability.
Stability of aggregates from plots with the association cassava-*Z.glabra* never correlated significantly with concentrations of organic matter and nutrients, indicating the harmful effect of soil disturbing activities on soil structure.
No significant changes occurred between treatments and locations from the first to the second cropping period except that aggregate stability increased in the second cropping period in Mondomo on cassava-vetiver grass plots (+1.85 mm), whereas it decreased in Quilichao (-0.70 mm).

Meanweight diameters at the two study sites were clearly different with 260 mm for Quilichao and 135 mm for Mondomo (averaged over both cropping periods) suggesting Quilichao soils to have a much greater stability. This, however, was not confirmed by empirically determined K-factors for the the two locations which indicated erodibility to be moderate to low at Quilichao (K = 0.015-0.019) and low at Mondomo (K = 0.011-0.012). The wet sieving procedure more closely approximates the slaking forces that are exerted on a soil by flowing runoff water rather than the forces of raindrop impact. Water stable aggregates resistant to breakdown by this method may not be stable when subjected to impacting water drops (Young, 1984).

After heavy rain events in Quilichao and Mondomo, surface crusts were observed on bare fallow plots and to a lesser degree on continuously cropped plots. Soil aggregates, overlying the crust and sorted by raindrop impact (and surface runoff), were sampled after two heavy rainstorms with similar USLE R-factors (Erosivity) on both locations to analyse their (dry) size distribution (data not shown): The proportion of aggregates > 2 mm from fallow plots was about twice as high in Mondomo with 37.1 % than in Quilichao with 18.5 %. This is in contradiction to results, obtained with the wet sieving method. In Quilichao, the proportion of water-stable aggregates > 2 mm was nearly three times as high with 33.0 % than in Mondomo. This may be an indication that determination of aggregate stability by raindrop impact (Castillo F., 1994; Bruce-Okine and Lal, 1975; McCalla, 1944) or a combination of both methods (Young, 1984; De Leenhheer and De Boodt, 1959) may be more appropiate to be applied on tropical Andean Inceptisols to determine susceptibility to erosion.

6.2.2 Particle size distribution

The method used for texture analysis of topsoils (0-20 cm; hydrometer) does not show the real particle size distribution, because organic matter is not previously destroyed, stable clay-humus particles remain intact.

As for chemical soil properties, plots cropped with cassava since 1986-87 did not differ significantly in their textural composition (Tables 23 and 25). Greatest effects were found for the date of sampling ($P<0.001$). Sand and silt contents increased and clay content decreased from May 1990 to April 1992. Increases in sand content were greater in Quilichao than in Mondomo. Silt content did not increase at Quilichao, when cassava was intercropped with *Z.glabra* (+0.4 %), but this cropping system was associated with a

3.4 % increase of silt at Mondomo. Greatest clay contents among cropping systems were found in this association in Quilichao, but smallest in Mondomo.
At the beginning of the study in May 1990, sand content was greatest and clay content smallest in newly cleared plots (Table 25). Bare fallow soil showed smallest silt contents, indicating a preference for removal of silt by the erosion process. Soils high in silt and very fine sand were found to be most susceptible to erosion in the corn belt of the USA (Wischmeier and Mannering, 1969); this particle size range together with the clay content are the factors to calculate the particle size parameter of the K- factor (erodibility) of the USLE.
Raindrops detach finer materials from surface aggregates, which are washed away and/or entrained in soil pores of the very permeable Inceptisols under study. Aggregates and coarser particles accumulate on the soil surface and are transported downslope by runoff. These rain sorted aggregates had 3.1 % greater sand contents than the bulk of the topsoil at Quilichao, whereas sand content was 3.5 % greater at Mondomo. This is probably a reason for the higher sand contents in sediments compared to the soil, where they erode from.

Table 25. Particle size distribution of topsoils (0-20 cm) in Quilichao and Mondomo in May 1990 and April 1992.

		QUILICHAO		MONDOMO	
		1990	1992	1990	1992
		%			
SAND	bare fallow	23.3	26.8	24.9	28.1
	old cropped	19.8	25.6	24.2	26.4
	new cropped	24.6	25.6	35.6	30.0
SILT	bare fallow	17.5	17.7	19.0	20.4
	old cropped	19.3	20.5	20.3	22.7
	new cropped	19.2	20.9	26.0	26.2
CLAY	bare fallow	59.2	55.5	56.1	51.5
	old cropped	60.9	53.9	55.5	50.9
	new cropped	56.2	53.5	38.4	43.8

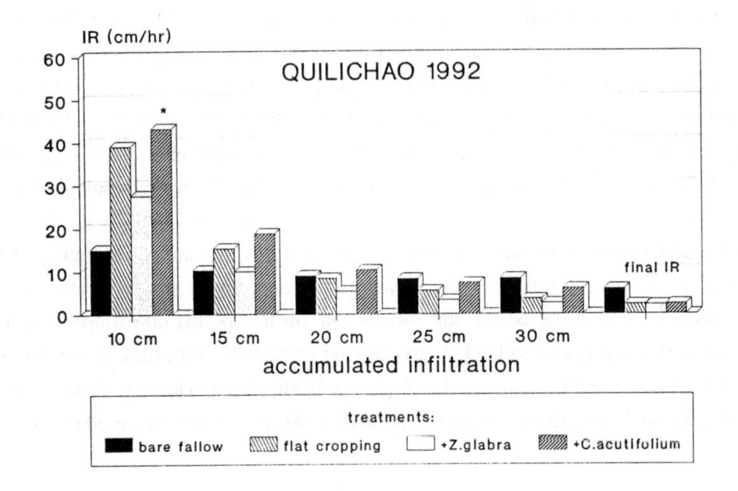

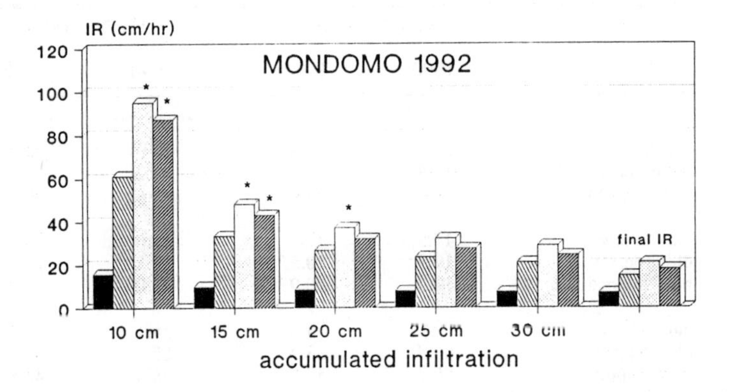

Figure 29: Infiltration rates for 10, 15, 20, 25 and 30 cm of accumulated infiltrated water and final rates at Quilichao and Mondomo by the end of the second cropping period 1992 in permanent bare fallow and selected cassava cropping systems.[a]

[a] * = significantly different from infiltration rate of bare fallow soil according to Dunnet's test.

Infiltration rates for accumulated infiltrated water amounting to 10, 15, 20, 25 and 30 cm and final rates of selected treatments are presented in Fig. 29 for the end of the second cropping period.
Final rates were high and generally exceed those given by Hillel (1971) for sandy soils. This is not surprising; soils are well aggregated and behave like sands through stable microaggregation (Ahn, 1979), in spite of their high clay contents.
Generally, infiltration itself and double-ring infiltrometer tests give highly variable results and require a large number of replicates to detect fine differences between cropping treatments, which was beyond the scope of this study. However, the following tendencies were observed:
Infiltration rates were higher on the steeper slopes in Mondomo than in Quilichao (Table 26). Significance decreased with increasing quantity of water infiltrated. Differences between selected treatments were highly significant at an early stage of infiltration; final rates, however, were not statistically different.

Table 26. Means and analysis of variance of infiltration rates (IR) at 10, 15, 20, 25, 30 cm accumulated infiltrated water and final rates in Quilichao and Mondomo 1992.

	Accumulated infiltration					
	10 cm	15 cm	20 cm	25 cm	30 cm	final IR
	cm h^{-1}					
Means						
Quilichao	31.2	13.6	8.3	6.1	5.2	3.2
Mondomo	64.9	33.6	26.1	22.8	20.5	15.2
	Analysis of variance					
TRT	***	**	*	+	ns	ns
LOC	***	**	**	**	*	*
TRTxLOC	*	*	*	**	+	*

Generally, significant interactions between locations and treatments occurred, suggesting that the same agronomic practices led to different infiltration rates depending on location, specific climate and soil properties. In Quilichao lowest permeability was found on permanent bare fallow plots in the beginning of the infiltration run. At 10 cm of infiltrated water, permeability was significantly greater when cassava was intercropped with *C.acutifolium*, as compared to the bare fallow treatment. Towards the end of the run, infiltration on bare plots was highest, but differences between treatments were not significant. The permeability of cropped treatments in Mondomo exceeded by far that in Quilichao. However, infiltration on bare fallow showed similar values over the whole run. This may be another indication, that at Quilichao the soils reacted with more sensitivity to soil disturbing operations, caused by trampling for the establishment of cassava and forage legumes, especially *Z.glabra*. At Quilichao, lower infiltration rates on plots with cassava and *Z.glabra* in Quilichao may explain the greater runoff in this treatment. At Mondomo, intercropping with forage legumes in Mondomo was associated with greater permeability than solecropped cassava, traditionally planted on the flat, and bare fallow. Differences to the fallow plots were significant during the early stages of infiltration. Water transmission at an early stage of infiltration (water absorption), is much lower on heavily eroded bare plots. This is attributed to the higher proportion of macropores in the topsoils of cropped plots, the influence of plant roots and soil cover by plants, weeds and mulch on aggregation and biotic activity. However, results have to be analysed with caution: Infiltration was measured in the dry season; soil moisture in bare plots can be assumed to be higher than in cropped plots, where soil water consumption is increased by transpiration demands. This results in drier soils and higher initial infiltration rates, whereas at later stages during the infiltration run, influence of subsoil permeability increases. Consequently, no significant differences between treatments were found for final rates.

The double-ring method does not account for the influence of soil crusts on conductivity. Crusts were formed on bare plots after rainfall. Cropped soils possess a greater aggregate stability than bare plots (Fig. 28), crust formation was rarely observed. Soil cover by cassava, weeds and mulch dissipates rainfall energy effectively from three to four months after crop establishment on. Crusts are characterized by a greater density, finer pores and a lower saturated conductivity than the underlying soil (Hillel, 1971). Under natural rainfall, water intake is diminished by soil crusts and can be assumed to be lower on bare plots than observed in Fig. 29, especially for the early phases of infiltration. This has important implications for the initiation of runoff and soil loss. Rain intensities may exceed frequently the infiltration rate on crusted and/or moist soils, especially in Quilichao, where final infiltration rates decreased to values below 30 $mm \cdot h^{-1}$. About 40 % of the rainfall in the study region fall with intensities > 25 $mm \cdot h^{-1}$.

6.2.4 Bulk density and soil water retention

Bulk densities and moisture retention properties of topsoils (0-20 cm) were determined in February/March 1991 (Table 27).

Bulk densities were lower in Mondomo (0.98 $g\ cm^{-3}$) than in Quilichao (1.04 $g\ cm^{-3}$). Differences between treatments were not significant. Slight interactions between locations and treatments occurred ($P < 0.1$). In Quilichao the densities of plots, which were under grass until June 1990 and since then cultivated with a cassava-vetiver grass system, showed some compaction, probably as a result of human stepping and cattle trampling during previous years. However, bulk densities were lowest in Mondomo on newly established plots (0.86 $g\ cm^{-3}$), previously under bush fallow (data not shown).

No significant differences among treatments occurred for the moisture retained at high ranges of matrix potential (-1 and -6 kPa), but for the lower ranges, differences were significant. The moisture content of soils with cassava and contour barriers of *P. purpureum* was higher from a potential of -6 kPa (pF=1.8) to -1500 kPa (pF=4.2). Significant differences were found at the permanent wilting point. However, these results are of limited value, since several soil factors render the determination of the moisture content at wilting point inaccurate. Taylor and Ashcroft (1972), state that soil moisture retention suffers modifications from temperature regime, degree of compaction, arrangement of pores in the soil.

Lal et al. (1979) found a significant improvement of moisture retention by cropping of legumes and grasses as planted fallows for two years on an eroded Alfisol, particularly at the high potential ranges. The authors attributed this to increased organic matter content and increased percentage of macropores in the fallow plots. Similar results were obtained by Hulugalle et al. (1986) with *Mucuna utilis* and Talineau et al. (1976) with *Stylosanthes guianensis* and *Centrosema pubescens*. In the present study, over all water potential ranges retention was lowest, when cassava was intercropped with legumes. Soil disturbance by human traffic for repeated establishment and cultivation operations (including cutting) and the time of measurements, less than one year after initiation of the trial, did not allow an improvement in macroporosity and organic matter content in these treatments.

Table 27. Bulk density, moisture retention at different potentials (kPa) and available moisture capacity of topsoils (0-20 cm) under bare fallow and selected cassava cropping systems in 1991(averaged over groth locations)..

Treatments	bulk density	moisture retention at different potentials (kPa)					available moisture capacity	
		-1	**-6**	**-30**	**-100**	**-1500**	**AMC1** [a]	**AMC2** [b]
	g cm^{-3}	Vol. %						
Bare fallow	1.02	50.6	44.6	41.4	38.6	32.2	12.4	9.2
Cassava cropping systems								
on the flat	1.02	51.3	45.0	39.3	37.3	33.5	11.5	5.8
+*Z.glabra*	1.00	50.3	44.1	38.5	34.9	30.8	13.3	7.7
+*C.acutifolium*	0.99	49.0	42.5	37.6	34.8	31.0	11.5	6.6
+*V.zizanioides*	1.00	50.3	45.4	41.2	39.3	35.6	9.8	5.6
+*P.purpureum*	1.07	51.3	45.9	41.5	39.9	36.0	9.9	5.5
LSD (0.05)	0.10	3.6	4.3	3.9	4.0	3.3	3.5	2.7
TRT	ns	ns	ns	ns	+	*	ns	ns
LOC	*	+	ns	+	*	ns	ns	ns
TRTxLOC	+	ns	ns	ns	ns	ns	ns	ns

[a] AMC1 = available moisture capacity between a matrix potential of -6 kPa and -1500 kPa.

[b] AMC2 = available moisture capacity between a matrix potential of -30 kPa and -1500 kPa.

High contents of clay, including amorphous materials, and of organic matter resulted in a high water retention of topsoils in the present study. Differences in contents of organic matter and clay between locations and treatments did not explain their variability in moisture retention. The effects of crop management practices may have been more important.

Available water capacity between -30 and -1500 kPa is low with levels between 5.5 and 9.2 Vol.%. Surprisingly, it was highest on heavily eroded bare fallow soils. The reason was a higher moisture retention between -100 and -1500 kPa of bare soils with 6.4 Vol.%, compared to 3.7 to 4.1 Vol.% of cropped plots (different at $P < 0.05$).

Probably the "real" available water capacity was somewhat higher because of volcanic constituents of soils in Quilichao and Mondomo. The field capacity of Andosols corresponds to a water potential of -10 kPa (Warkentin and Maeda, 1980).

Limited data (data not shown) obtained one year after the first measurements indicated an increase in moisture retention at a potential of -1 kPa in topsoils, highest in cassava-forage

legume intercropping systems compared to sole flat cropping. When cassava was intercropped with *C.acutifolium* water retention at -1 kPa was 7.6 Vol.% greater in Quilichao and 4.2 Vol.% greater in Mondomo than the year before. The differences between years were significant at $P < 0.05$ for the intercropping treatments only.
Available moisture (between -30 and -1500 kPa) did not change in Quilichao, when cassava was flat cropped or associated with *Z.glabra*, but increased, when intercropped with *P.phaseoloides* (+ 3.7 Vol.%; $P < 0.05$) and *C.acutifolium* (+ 3.4 Vol.%; $P < 0.05$). In Mondomo, no significant changes ocurred.

Water retention at the lower range of matrix potential depends primarily on structural characteristics, retention at the permanent wilting point on the particle size distribution (El-Swaify, 1980; Manrique et al. 1991). Bulk density was inversely correlated with moisture retention at -1 kPa ($r = -0.77$; $P < 0.001$) and at -6 kPa ($r = -0.53$; $P < 0.05$). However, no significant correlation was found between -1500 kPa and any textural group.

6.3 *Conclusions*

The number of years under cultivation had the greatest effect on soil chemical properties. Cropping decreased organic matter content and increased soil acidity compared to original soil under grassland (Quilichao) and bush fallow (Mondomo). Levels of available nutrients in topsoils, removed by harvest and lost by soil erosion, were restored (available K) or steadily improved (Bray-II P, exchangeable Ca and Mg) by annual fertilization. Differences among cropping systems were generally not significant. Probable reasons are the short trial period of two years and the way the cropping systems were managed. Additional nutrient loss on intercropped plots by the removal of the fodder plants was probably compensated by the lower extraction of nutrients with cassava, which suffered strong growth and yield reductions through intercropping with legumes.
Physical properties of soils were favorable. Contour ridging increased water-stable aggregation to levels similar to long-term grass or bush fallow, which may be one reason for the good cassava yields in this treatment. Intensities of rainfall easily exceeded water infiltration rates, especially on heavily eroded and crusted soils, leading to high runoff and often soil losses. In Quilichao, the soil was probably more susceptible to "human traffic" than in Mondomo. In Quilichao, the smallest water permeability and greatest surface runoff

among cropping treatments was found in the cassava-*Z. glabra* association, where the intensity of "human traffic" for the legume establishment was greatest. Limited data showed an increase in moisture retention at high matrix potentials by forage legumes such as *C. acutifolium*. Available moisture showed a tendency to increase by intercropping of cassava with *C. acutifolium* and *P. phaseoloides*. Probably, forage legumes, cut and left in the plots and grown for a longer time period, would have had greater positive effects on soil fertility and soil structure in cassava cropping.

Inceptisols in the montaneous regions of South West Colombia are of low soil fertility, but of an excellent physical stability. Eroded hillsides, once soil fertility constraints can be alleviated, still possess the potential of agricultural production.

Part 7 *PRACTICAL CONCLUSIONS AND RESEARCH SUGGESTIONS*

Sound crop and soil management such as the use of fertilizers and lime, pest and disease control, optimum planting densities and patterns, selection of appropriate crop varieties among others are fundamental to diminish soil erosion and achieve greater yields. From Reining's study (Reining, 1992) and the present study, it may be concluded, that water erosion on South West Colombian Inceptisols can be controlled effectively and acceptable yields can be achieved by **(i)** cropping of cassava on properly constructed contour ridges on slopes up to 20 % and **(ii)** using contour strips of densly planted grass barriers at adequate distances from 5 % slope on.

Forage legumes such as *P.phaseoloides*, *Z.glabra*, *C.acutifolium* and *C.macrocarpum*, simultaneously established with cassava and used as companion crops did not provide a fast soil cover in the first three months of cassava cropping, when the erosion risk was greatest. In Quilichao, additional walking on the plots for legume establishment enhanced soil losses significantly compared to traditional cassava sole cropping. When cassava was established by minimum tillage ("cajuelas") in the existing legume swards soil erosion was controlled effectively only if dense stands of legumes provided an uniformly distributed soil cover. However, cassava yields were greatly reduced (> 40%). Smaller yields may be attributed to competition from the legumes and the negative effect of reduced tillage on these soils.

Loss of cropping area by grass contour strips could be compensated to some extent by increasing the cassava planting density in the remaining area. By choosing appropriate grass species and distances between the first cassava row and the strip, yield reduction of cassava through competition from grasses may be minimized.

Farmers in the study region are not likely to adopt erosion control practices, if no direct benefit is obvious. Therefore, an urgent need does not only exist to identify plant species suitable for erosion control, but also to increase crop yields and crop diversity. Plant species should be easy to establish on already degraded soils and develop a fast soil cover, be not too competitive with the main crop and, above all, provide direct or indirect income (fodder, construction materials etc.) to the farmer. Effective erosion control practices for cropping on steeper slopes, than those used in the present study, should be developed. The great potential of reduced tillage systems and mulch application for erosion control needs further clarification in relation to crop productivity and adaptability to the site specific socio-economic conditions in the study region.

REFERENCES

Adimihardja, A. 1989. Rainfall erosivity and soil erodibility in Indonesia: Estimation and variation with time. Ph.D. Thesis, Faculty of Agricultural Sciences, Rijksuniversiteit Ghent, Belgium.

Ahmad, N., and E. Breckner. 1974. Soil erosion on three Tobago soils. Tropical Agriculture (Trinidad) 51:313-324.

Ahn, P.M. 1979. Microaggregation in tropical soils: Its measurement and effects on the maintenance of soil productivity. *In*: R. Lal and D.J. Greenland (ed.) Soil Physical Properties and Crop Production in the Tropics, pp. 75-85. John Wiley, Chichester, UK.

Aina, P.O. 1979. Soil changes resulting from long term management practices in Western Nigeria. Soil Science Society of America. Journal 43: 173-177.

Aina, P.O., R. Lal, and G.S. Taylor. 1976. Soil and crop management in relation to soil erosion in the rainforest region of Western Nigeria. Symposium Proceedings. National Soil Erosion Conference, May 25-26, 1976. Lafayette, Indiana, USA.

Aina, P.O., R. Lal, and G.S. Taylor. 1979. Effects of vegetal cover on soil erosion on an Alfisol. *In*: R. Lal and D.J. Greenland (ed.) Soil Physical Properties and Crop Production in the Tropics, pp. 501-507. John Wiley, Chichester, UK.

Alberts, E.E., and W.C. Moldenhauer. 1981. Nitrogen and phosphorus transported by eroded soil aggregates. Soil Science Society of America. Journal 45: 391-396.

Alberts, E.E., G.E. Schuman, and R.E. Burwell. 1978. Seasonal runoff losses of nitrogen and phosphorus from Missouri valley loess watersheds. Journal of Environmental Quality 7: 203-208.

Alberts, E.E., W.C. Moldenhauer, and G.R. Foster. 1980. Soil aggregates and primary particles transported in rill and interrill flow. Soil Science Society of America. Journal 44: 590-595.

Barisas, S.G., J.L. Baker, H.P. Johnson, and J.M. Laflen. 1978. Effect of tillage systems on runoff losses of nutrients, a rainfall simulation study. Transactions. American Society of Agricultural Engineers 21: 893-897.

Barnett, A.P., J.R. Carreker, F. Abruna, and A.E. Dooley. 1971. Erodibility of selected tropical soils. Transactions. American Society of Agricultural Engineers 14: 496-499.

Barrows, H.L., and V.J. Kilmer. 1963. Plant nutrient losses from soils by water erosion. Advances in Agronomy 15: 303-316.

Bharad, G.M., and B.C. Bathkal. 1990. Role of vetiver grass in soil and moisture conservation. *In*: Proceedings of the Colloquium on the Use of Vetiver in Sediment Control. April 25, 1990. Watershed Management Directorate, Dehra Dun, India.

Bomah, A.K. 1988. Rainfall conditions and erosivity in the Nyala Area of Sierra Leone. Journal of Environmental Management 26: 1-7.

Bouwer, H. 1986. Intake rate: Cylinder infiltrometer. *In*: A. Klute (ed.) Methods of soil analysis. Part 1: Physical and mineralogical methods, pp. 825-844. 2nd Ed., American Society of Agronomy. Madison, WC, USA.

Bremner, J.M., and C.S. Mulvaney. 1982. Nitrogen-Total. *In*: A.L. Page, R.H. Miller, and D.R. Keeney (ed.) Methods of soil analysis. Part 2. Chemical and microbiological properties, pp. 595-624 . American Society of Agronomy. Madison, WC, USA.

Bruce-Okine, E., and R. Lal. 1975. Soil erodibility as determined by raindrop technique. Soil Science 119: 149-157.

Bryan, R.B. 1968. The development, use and efficiency of indices of soil erodibility. Geoderma 2: 5-25.

Burwell, R.E., D.R. Timmons, and R.F. Holt. 1975. Nutrient transport in surface runoff as influenced by soil cover and seasonal periods. Soil Science Society of America. Proceedings 39: 523-528.

Cadavid L.F., and R.H. Howeler. 1987. El problema de la erosión en los suelos de Mondomo, Cauca, Colombia, dedicados al cultivo de la yuca y sus posibles soluciones. Centro Internacional de Agricultura Tropical (CIAT), Cali, Colombia.

Castillo F., J.A. 1994. Determinación del indice de erodabilidad (K) en dos suelos del Cauca, Colombia. 206 p. M.Sc. Thesis. Universidad Nacional de Colombia. Posgrado Suelos y Aguas, Sede Palmira, Colombia.

Chaney, K., and R.S. Swift. 1984. The influence of organic matter on aggregate stability in some British soils. Journal of Soil Science 35: 223-230.

Centro Internacional de Agricultura Tropical (CIAT). 1978. Annual report for 1977. Cali, Colombia.

Centro Internacional de Agricultura Tropical (CIAT). 1979. Annual report for 1978. Cali, Colombia.

Centro Internacional de Agricultura Tropical (CIAT). 1984. Annual Report for 1982 and 1983. Cassava Program. Cali, Colombia.

Centro Internacional de Agricultura Tropical (CIAT). 1988. Annual Report for 1987. Tropical Pastures Program. Working Document No. 44. Cali, Colombia.

Centro Internacional de Agricultura Tropical (CIAT). 1991a. Annual Report for 1990. Cassava Program. Working Document No.95. Cali, Colombia.

Centro Internacional de Agricultura Tropical (CIAT). 1991b. Annual Report, for December 1990. Cassava Program. For internal circulation and discussion only. pp. 225,226,232. Cali, Colombia

Centro Internacional de Agricultura Tropical (CIAT). 1994. Annual Report for 1993. Cassava Program. Cali, Colombia.

Cock, J.H. 1985. Cassava. New potential for a neglected crop. 192p. Westview Press Inc., Boulder, Colorado.

Connor, D.J., J.H. Cock, and G.E. Parra. 1981. Response of cassava to water shortage. I. Growth and yield. Field Crops Research 4: 181-200.

Dangler, E.W., and S.A. El-Swaify. 1976. Erosion of selected Hawaii soils by simulated rainfall. Soil Science Society of America. Journal 40: 769-773.

Day, P.R. 1965. Particle fractionation and particle size analysis. *In*: C.A. Black (ed.) Methods of soil analysis. Part 1, pp. 545-567. Agronomy Monograph No. 9. American Society of Agronomy. Madison, WI, USA.

De Leenheer, L., and M. De Boodt. 1959. Determination of aggregate stability by the change in mean weight diameter. International Symposium Soil Structure. Mededelingen. Landbouwhogeschool (Ghent) 24: 290-300.

Eck, H.V. 1968. Effect of topsoil removal on nitrogen supplying ability of Pullman silty clay loam. Soil Science Society of America. Proceedings 32: 686-691.

Eck, H.V. 1969. Restoring productivity on Pullman silty clay loam subsoil under limited soil moisture. Soil Science Society of America. Proceedings 33: 578-581.

Egashira, K., Y. Kaetsu, and K. Takuma. 1983. Aggregate stability as an index of erodibility of Ando Soils. Soil Science and Plant Nutrition 29: 473-481.

Egashira, K., S. Nakai, and K. Takuma. 1986. Relation between soil properties and erodibility of Red-Yellow (Ultisols) B soils. Soil Science and Plant Nutrition 32: 551-559.

El-Swaify, S.A. 1977. Susceptibilities of certain tropical soils to erosion by water. *In*: D.J. Greenland and R. Lal (ed.) Soil conservation and management in the humid tropics, pp. 71-77. John Wiley, Chichester, Untided Kingdom.

El-Swaify, S.A. 1980. Physical and mechanical properties of Oxisols. *In*: B.K.G. Theng (ed.) Soils with Variable Charge, pp. 372-324. New Zealand Society of Soil Science, Soil Bureau, Department of Scientific and Industrial Research, Lower Hutt, New Zealand.

El-Swaify, S.A. 1991. Land-based limitations and threats to world food production. Outlook on Agriculture 20: 235-242.

El-Swaify, S.A., and E.W. Dangler. 1976. Erodibilities of selected tropical soils in relation to structural and hydrological parameters. *In*: Soil Erosion: Prediction and Control, pp. 105-114. Soil and Water Conservation Society. Ankeny, IA, USA.

El-Swaify, S.A., S. Arunin, and I.P. Abrol. 1983. Soil Salinization: Development of salt affected soils. *In*: R.A. Carpenter (ed.) Natural Systems for Development, What Planners Need to Know, pp. 162-228. Macmillan, New York, USA.

El-Swaify, S.A., A. Lo, R. Joy, L. Shinshiro, and R.S. Yost. 1988. Achieving conservation-effectiveness in the tropics using legume intercrops. Soil Technology 1: 1-12.

Elwell, H.A., and M.A. Stocking. 1973. Rainfall parameters for soil loss estimation in a subtropical climate. Journal of Agricultural Engineering Research 18: 169-177.

Epstein E., W.J. Grant, and R.A. Struchtemeyer. 1966. Effects of stones on runoff, erosion, and soil moisture. Soil Science Society of America. Proceedings 30: 638-640.

Foster, G.R. 1982. Modelling the erosion process. *In*: C.T. Haan, H.P. Johnson and D.L. Brakensiek (ed.) Hydrologic modelling of small watersheds, pp. 297-382. American Society of Agricultural Engineers. St.Joseph, MI, USA.

Foster, G.R., and L.D. Meyer. 1975. Mathematical simulation of upland erosion using fundamental erosion mechanics. Proceedings. Sediment Yield Workshop, pp. 190-207. United States Department of Agriculture. Sedimentation Laboratory, Oxford, MS ARS-S-40, USA.

Foster, G.R., D.K. McCool, K.G. Renard, and W.C. Moldenhauer. 1981. Conversion of the universal soil loss equation to SI metric units. Journal of Soil and Water Conservation 36: 355-359.

Foster, G.R., F. Lombardi, and W.C. Moldenhauer. 1982. Evaluation of a rainfall-runoff erosivity factor for individual storms. Transactions. American Society of Agricultural Engineers 25: 124-129.

Foster, G.R., R.A. Young, Römkens, M.J.M., and C.A. Onstad. 1985. Processes of soil erosion by water. *In*: R.F. Follett and B.A. Stewart (ed.) Soil erosion and crop productivity, pp. 137-162. American Society of Agronomy-Crop Science Society of America-Soil Science Society of America. Madison, WI 53711,USA.

Fournier, F. 1967. Research on soil erosion and conservation in Africa. Sols Africains 12: 53-96.

Franco F., H., and M. Gonzalez, A. 1967. Comparacion de algunos métodos para determinar la estabilidad de los agregados al agua. Acta Agronomica (Colombia) 17: 21-41.

Frye, W.W., S.A. Ebelhar, L.W. Murdock, and R.L. Blevins. 1982. Soil erosion effects on properties and productivity of two Kentucky soils. Soil Science Society of America. Journal 46: 1051-1055.

Gee, G.W., and J.W. Bauder 1986. Particle-size analysis. *In*: A. Klute (ed.) Methods of Soil Analysis, Part 1. Physical and Mineralogical Methods, pp. 383-411. Agronomy Monograph no.9 (2nd Edition). American Society of Agronomy-Soil Science Society of America. Madison, WI, USA.

Gumbs, F.A., J.I. Lindsay, M. Nasir, and Angella Mohammed. 1985. Soil erosion studies in the northern mountain range, Trinidad, under different crop and soil management. *In*: S.A. El-Swaify, W.C. Moldenhauer and A. Lo (ed.) Soil erosion and conservation, pp. 90-98. Soil Conservation Society of America. Ankeny, IA, USA.

Hegewald, H.B. 1990. Screening of different tropical legumes in monoculture and in association with cassava for adaptation to acid infertile and high Al content soil. Beiträge zur tropischen Landwirtschaft und Veterinärmedizin 28: 283-289.

Hillel, D. 1971. Soil and Water. Physical Principles and Processes, pp.131-153. Academic Press, New York, USA.

Howeler, R.H. 1985a. Practicas de conservacion de suelos para cultivos anuales. *In*: R.H. Howeler (ed.) Manejo y conservacion de suelos. Memorias del primer seminario sobre manejo y conservacion de suelos, pp. 77-93. Junio 14 - 16, 1984, Cali, Colombia.

Howeler, R.H. 1985b. Potassium nutrition of cassava. *In*: R.O. Munson (ed.) Potassium in Agriculture, pp. 819-841. American Society of Agronomy-Crop Science Society of America-Soil Science Society of America. Madison, WI, USA.

Hudson, N.W. 1961. An introduction to the mechanics of soil erosion under conditions of subtropical rainfall. Rhodesian Science Assessment Proceedings and Transactions XLIX: 14-25.

Hudson, N.W. 1971. Soil conservation. 320 p. Batsford Ltd. London, UK.

Hulugalle, N.R., R. Lal, and C.H.H. Ter Kuile. 1986. Amelioration of soil physical properties by Mucuna after mechanized land clearing of a tropical rain forest. Soil Science 141: 219-224.

Hulugalle, N.R., and H.C. Ezumah. 1991. Effects of cassava-based cropping systems on physico-chemical properties of soil and earthworm casts in a tropical Alfisol. Agriculture, Ecosystems and Environment 35: 55-63.

IGAC. 1976. Estudio general de suelos de los municipios Santander de Quilichao, Piendamo, Morales, Buenos Aires, Cajibio y Caldono (Departamento del Cauca). 472p. Instituto Geografico Augustin Codazzi. Vol XII (4). Bogotá, Colombia.

Jansson, M.B. 1982. Land erosion by water in different climates. 151 p. UNGI Rapport Nr. 57. Uppsala, Sweden.

Keeney, D.R., and O.W. Nelson. 1982. Nitrogen-Inorganic Forms. *In*: A.L. Page, R.H. Miller, and D.R. Keeney (ed.) Methods of soil analysis. Part 2: Chemical and microbiological properties, pp. 643-698. American Society of Agronomy-Soil Science Society of America. Madison, WI, USA.

Kemper, W.D., and E.J. Koch. 1966. Aggregate stability of soils from Western United States and Canada. United States Department of Agriculture. Technical Bulletin No. 1355.

Kemper, W.D., and R.C. Rosenau. 1986. Aggregate stability and size distribution. *In*: A. Klute (ed.) Methods of Soil Analysis, Part 1. Physical and Mineralogical Methods, pp. 425-443. Agronomy Monograph no.9 (2nd Edition). American Society of Agronomy-Soil Science Society of America. Madison, WI, USA.

Kinnel, P.I.A. 1973. The problem assessing the erosive power of rainfall from meteorological observations. Soil Science Society of America. Proceedings 37:617- 621.

Knisel, W.G., and G.R. Foster. 1981. CREAMS: A system for evaluating best management practices, pp 177-194. *In*: Economics, ethics, ecology: Roots of productive conservation. Soil Conservation Society of America. Ankeny, IA, USA.

Kowal, J.M., and A.H. Kassam. 1976. Energy load and instantaneous intensity of rainstorms at Samaru, Northern Nigeria. Tropical Agriculture (Trinidad) 53: 185-197.

Lal, R. 1976a. Soil erosion problems on an Alfisol in western Nigeria and their control. International Institute of tropical Agriculture (IITA). Monograph No. 1. Ibadan, Nigeria.

Lal, R. 1976b. No-tillage effects on soil properties under different crops in Western Nigeria. Soil Science Society of America. Journal 40: 762-768.

Lal, R. 1976c. Soil erosion problems on Alfisols in Western Nigeria. IV. Nutrient element losses in runoff and eroded sediments. Geoderma 16: 403-417.

Lal, R. 1976d. Soil erosion on Alfisols in Western Nigeria. V. The changes in physical properties and the responses of crops. Geoderma 16: 419-431.

Lal, R. 1979. Physical characteristics of soils of the tropics: Determination and management. *In*: R. Lal and D.J. Greenland (ed.) Soil Physical Properties and Crop Production in the Tropics, pp. 7-44. John Wiley, Chichester, UK.

Lal, R. 1980. Soil erosion as a constraint to crop production. *In*: Soil related constraints to food production in the tropics, pp. 405-423. International Rice Research Institute (IRRI), Los Baños, Laguna, Philippines.

Lal, R. 1985. Soil erosion and its relation to productivity of tropical soils. *In*: S.A. El-Swaify, W.C. Moldenhauer and A. Lo (ed.) Soil Erosion and Conservation, pp. 237-247. Soil Conservation Society of America, Ankeny, IA, USA.

Lal, R. 1987. Tropical Ecology and Physical Edaphology. pp.135-150. John Wiley, Chichester, UK.

Lal, R, G.F. Wilson, and B.N. Okigbo. 1979. Changes in properties of an Alfisol produced by various crop covers. Soil Science 127: 377-382.

Langdale, G.W., J.E. Box Jr., R.A. Leonard, and A.P. Barnett. 1979. Corn yield reduction on eroded southern Piedmont soils. Journal of Soil and Water Conservation 34: 226-228.

Laws, J.O., and D.A. Parsons. 1943. The relationship of raindrop size to intensity. Transactions. American Geophysical Union 24:452-460.

Lehle, M. 1986. Erosionsuntersuchungen im Nährpflanzenanbau auf andinen Hangflächen Kolumbiens. 93p. M.Sc. Thesis, unpublished. Institut für Pflanzenproduktion in den Tropen und Subtropen, Universität Hohenheim, Stuttgart, Germany.

Lo, A., S.A. El-Swaify, E.W. Dangler, and L. Shinshiro. 1985. Effectiveness of EI_{30} as an erosivity index in Hawaii. *In*: El-Swaify, W.C. Moldenhauer and A. Lo (ed.) Soil erosion and conservation, pp.384-392.S.A.. Soil Conservation Society of America, Ankeny, IA, USA.

Lugo-Lopez, M.A., J. Suarez Jr., and J.A. Bonnet. 1968. Relative infiltration rate of Puerto Rican soils. Journal of the Agricultural University of Puerto Rico 52: 233-240.

Luk, S.H. 1979. Effect of soil properties on erosion by wash and splash. Earth Surface Processes 4: 241-255.

Maass, J.M., C.F. Jordan, and J. Sarukhan. 1988. Soil erosion and nutrient losses in seasonal tropical agroecosystems under various management techniques. Journal of Applied Ecology 25: 595-607.

Manrique, L.A., C.A. Jones, and P.T. Dyke. 1991. Predicting soil water retention characteristics from soil physical and chemical properties. Communications in Soil Science and Plant Analysis 22: 1847-1860.

Margolis, E., I.C. Galindo de L., and A.V. de Mello-Netto. 1991. Comportamento de sistemas de cultivo de mandioca em relacao a produccao e as perdas por erosao. Revista Brasileira de Ciencia do Solo 15: 357- 362.

Massey, H.F., and M.L. Jackson. 1952. Selective erosion of fertility constituents. Soil Science Society of America. Proceedings 16: 353-356.

Mbagwu, J.S.C., R. Lal, and T.W. Scott. 1984. Effect of artificial desurfacing of Alfisols and Ultisols in Southern Nigeria. 1. Crop performance. Soil Science Society of America. Journal 48: 828-831.

McCalla, T.M. 1944. Water drop method of determining stability of soil structure. Soil Science 58: 117-121.

McDowell, L.L., and K.C. McGregor. 1984. Plant nutrient losses in runoff from conservation tillage corn. Soil and Tillage Research 4: 79-91.

Metha, N.C., J.D. Legg, C.A.I. Goring, and C.A. Black. 1954. Determination of organic phosophorus in soils. I. Extraction methods. Soil Science Society of America. Proceedings 18: 443-449.

Meyer, L.D., D.E. Line, and W.C. Harmon. 1992. Size characteristics of sediment from agricultural soils. Journal of Soil and Water Conservation 47: 107-111.

Muhr, L., D.E. Leihner, T.H. Hilger, K.M. Müller-Sämann. 1995. Intercropping of cassava with herbaceous legumes: I Rooting patterns and their below-ground competition. Angewandte Botanik 69 (in press).

Mulongoy, K., and I.O. Akobundu. 1985. Nitrogen uptake in live mulch systems. *In*: B.T Kang, J. van der Heide (ed.) Nitrogen management in farming systems in humid and subhumid tropics, pp. 285-290. Institute for Soil Fertility. Haren, The Netherlands.

Munn, D.A., E.O. Mc Lean, A. Ramirez, and T.J. Logan. 1973. Effect of soil, cover, slope, and rainfall factors on soil and phosphorus movement under simulated rainfall conditions. Soil Science Society of America. Proceedings 37: 428-431.

Mutchler, C.K., and C.E. Carter. 1983. Soil erodibility variation during the year. Transactions. American Society of Agricultural Engineers 26: 1102-1104, 1108.

Nitis I.M., and I.G.N. Sumatra. 1976. The effect of fertilizers on the growth and yield of Cassava (Manihot esculenta var. gading) undersown with Stylo (Stylosanthes guyanensis cv. Schofield) at Penebel, Bali. 13p. Bulletin Fakultas Kedokteran Hewan&Peternakan. Universitas Udayana. Denpasar, Bali, Indonesia.

Oades, J.M. 1984. Soil organic matter and structural stability: mechanisms and implications for management. Plant and Soil 76: 319-337.

Olson, S.R., and L.E. Sommers. 1982. Phosphorus. *In*: A.L. Page, R.H. Miller, and D.R. Keeney (ed.) Methods of soil analysis. Part 2: Chemical and microbiological properties, pp. 403-430. American Society of Agronomy. Madison, WI, USA.

Paez, M.L., and O.S. Rodriguez. 1989. Factores de la Ecuacion Universal de Perdidas de Suelo in Venezuela. *In*: M.L. Paez (ed.) Conservacion de Suelos y Aguas. La Erosion Hydrica, Diagnostico y Control. Revista de la Facultad de Agronomia de la Universidad Central de Venezuela, Alcance No. 37, pp. 21-31. Maracay, Venezuela.

Reddy, G.Y, E.O. Mc Lean, G.D. Hoyt, and T.J. Logan. 1978. Effects of soil, cover crop, and nutrient source on amounts and forms of phosphorus movement under simulated rainfall conditions. Journal of Environmental Quality 7: 50-54.

Reining, L. 1992. Erosion in Andean Hillside Farming. 219 p. Hohenheim Tropical Agricultural Series No.1. Verlag Josef Margraf, Weikersheim, Germany.

Richards, L.A. 1965. Physical condition of water in soil. *In*: C.A. Black (ed.) Methods of Soil Analysis. Part 1, pp. 128-152. American Society of Agronomy. Madison, WI, USA.

Rixon, A.J. 1966. Soil fertility changes in a red-brown earth under irrigated pastures. 1. Changes in organic carbon, carbon/nitrogen ratio, cation exchange capacity and pH. Australian Journal of Agricultural Research 17: 303-316.

Röhmkens, M.J.M. 1985. The soil erodibility factor: A perspective. *In*: S.A. El-Swaify, W.C. Moldenhauer and A. Lo (ed.) Soil erosion and Conservation, pp. 445-461. Soil Conservation Society of America. Ankeny, IA, USA.

Röhmkens, M.J.M., D.W. Nelson, and C.B. Roth. 1975. Soil erosion on selected high clay subsoils. Journal of Soil and Water Conservation 30: 173-176.

Roose, E.J. 1973. Dix-sept annees de mesures experimentales de l'erosion et du ruissellement au Senegal. Agronomie Tropicale 22: 123-152.

Roose, E.J. 1977. Use of the Universal Soil Loss Equation to predict erosion in West Africa. *In*: Soil erosion: Prediction and control, pp. 60-74. Soil Conservation Society of America, Special Publication No. 21. Ankeny, IA, USA.

Roose, E.J. 1980. Approach to the definition of rain erosivity and soil erodibility in West Africa. *In*: M. De Boodt and D. Gabriels (ed.) Assessment of erosion. John Wiley, Chichester, UK.

Ryan, K.T. 1986. Soil conservation research summary. Thai-Australia-World Bank Land Development Project.

Sabel-Koschella, U. 1988. Field studies on soil erosion in the Southern Guinea Savanna of Western Nigeria. Ph.D. Thesis, Institut für Bodenkunde der Technischen Universität München in Weihenstephan, Germany.

Salinas, J.G., and R. Garcia. 1985. Métodos químicos para el analysis de suelos acidos y plantas forrajeras. 83p. CIAT, Cali, Colombia.

SAS. 1988. SAS/STAT^{IM} User's Guide, Release 6.03 Edition. 1028p. SAS Institute Inc.. Cary, NC, USA.

Sethi, K.L. 1982. Breeding and cultivation of new Khas hybrid clones. Indian Perfumer 26 (2-4): 54-61.

Sharpley, A.N. 1980. The enrichment of soil phosphorus in runoff sediments. Journal of Environmental Quality 9: 521-526.

Sharpley, A.N. 1985. The selective erosion of plant nutrients in runoff. Soil Science Society of America. Journal 49: 1527-1534.

Stocking, M.A. 1988. Assessing vegetative cover and management effects. *In*: R. Lal (ed.) Soil Erosion Research Methods, pp. 163-185. Soil and Water Conservation Society, Ankeny, IA, USA.

Stocking, M.A., and H.A. Elwell. 1973. Prediction of subtropical storm soil losses from field plot studies. Agricultural Meteorology 12: 193-201.

Stocking, M., and L. Peake. 1986. Crop yield losses from the erosion of Alfisols. Tropical Agriculture (Trinidad) 63: 41-45.

Talineau, J.C., G. Hainnaus, B. Bonzon, C. Eillornneau, D. Picard, and M. Sicot. 1976. Agronomic aspects of the inclusion of a forage crop in a crop rotation under humid tropical conditions in the Ivory coast. Cahiens. Office de la Recherche Scientifique et Technique d'Outre Mer (ORSTROM), Serie Biologie 11: 277-290.

Taylor, S.A., and G.L. Ashcroft. 1972. Physical Edaphology: The Physics of Irrigated and Nonirrigated Soils. 533p.. W.H. Freemann and Company, San Francisco, USA.

Timmons, D.R., R.F. Holt, and J.J. Latterell. 1970. Leaching of crop residues as a source of nutrients in surface runoff water. Water Resources Research 6: 1367-1375.

Timmons, D.R., R.E. Burwell, and R.F. Holt. 1973. Nitrogen and phosphorus losses in surface runoff from agricultural land as influenced by placement of broadcast fertilizer. Water Resources Research 9: 658-667.

Tisdall, J.M., and J.M. Oades. 1982. Organic matter and water-stable aggregates in soils. Journal of Soil Science 33: 141-163.

Trott, K.E., and M.J. Singer. 1983. Relative erodibility of 20 California range and forest soils. Soil Science Society of America. Journal 47: 753-759.

Tscherning, K., D.E. Leihner, T.H. Hilger, K.M. Müller-Sämann, and M.A. El Sharkawy. 1995. Grass barriers in cassava hillside cultivation - rooting patterns and root growth dynamics. Field Crops Research (accepted for publication).

Tukey Jr., H.B., H.B. Tukey, and S.H. Wittwer. 1958. Loss of nutrients by foliar leaching as determined by radioisotopes. Proceedings. American Society of Horticultural Science 71: 496-506.

Ulsaker, L.G., and C.A. Onstad. 1984. Relating rainfall erosivity factors to soil loss in Kenya. Soil Science Society of America. Journal 48: 891-896.
USDA. 1951. Soil Survey manual. Agricultural Handbook No. 18. United States Department of Agriculture. Washington, D.C., USA.

Van Bavel, C.H.M. 1949. Mean weight diameter of soil aggregates as a statistical index of aggregation. Soil Science Society of America. Proceedings 14: 20-23.

Warkentin, B.P., and T. Maeda. 1980. Physical and mechanical characteristics of Andisols. *In*: B.K.G. Theng (ed.) Soils with Variable Charge, pp. 281-302. New Zealand Society of Soil Science, Soil Bureau, Department of Scientific and Industrial Research, Lower Hutt, New Zealand.

Wilaipon, B., R.C. Gutteridge, and K. Chutikul. 1981. Undersowing upland crops with pasture legumes. 1. Cassava with Stylosanthes hamata cv Verano. Thai Journal of Agricultural Science 14: 333-337.

Wilkinson, G.E., and P.O. Aina. 1976. Infiltration of water into two Nigerian soils under secondary forest and subsequent arable cropping. Geoderma 15: 51-59.

Williams, C.H., and C.M. Donald. 1957. Changes in organic matter and pH in a podzolic soil as influenced by subterranean clover and superphosphate. Australian Journal of Agricultural Research 8: 179-189.

Williams, J.R., R.R. Allmaras, K.G. Renard, L. Lyles, W.C. Moldenhauer, G.W. Langdale, L.D. Meyer, W.J. Rawls, G. Darby, R. Daniels, and R. Magleby. 1981. Soil erosion effects on productivity: A research perspective. Journal of Soil and Water Conservation 36: 82-90.

Wischmeier, W.H., and D.D. Smith. 1958. Rainfall energy and its relation to soil loss. Transactions. American Geophysical Union 39: 285-291.

Wischmeier, W.H. 1959. A rainfall erosion index for a Universal Soil Loss Equation. Soil Science Society of America. Proceedings 23: 246-249.

Wischmeier, W.H. 1960. Cropping-management factor evaluations for a universal soil-loss equation. Soil Science Society of America. Proceedings 23: 322-326.

Wischmeier, W.H, and J.V. Mannering. 1969. Relation of soil properties to its erodibility. Soil Science Society of America. Proceedings 33: 131-137.

Wischmeier, W.H., C.B. Johnson, and B.V. Cross. 1971. A soil erodibility nomograph for farmland and construction sites. Journal of Soil and Water Conservation 26: 189-193.

Wischmeier, W.H, and D.D. Smith. 1978. Predicting rainfall erosion losses - a guide to conservation planning. 58p. Agricultural Handbook No. 537. United States Department of Agriculture. Washington, D.C., USA.

Wolf, J.M., and M. Drosdoff. 1976. Soil water studies in Oxisols and Ultisols of Puerto Rico: I. Water movement. Journal of the Agricultural University of Puerto Rico 60: 375-385.

Young, R.A. 1984. A method of measuring aggregate stability under waterdrop impact. Transctions. American Society of Agricultural Engineers 27: 1351-1354.

Young, R.A., and C.K. Mutchler. 1977. Erodibility of some Minnesota soils. Journal of Soil and Water Conservation 32: 180-182.

Young, R.A., A.E. Olness, C.K. Mutchler, and W.C. Moldenhauer. 1986. Chemical and physical enrichments of sediment from cropland. Transactions. American Society of Agricultural Engineers 29: 165-169.